"This text provides an integrative, (very!) specific Christian digital sexual ethics. It expands our Christian moral imaginations, equipping us for the current digital age and what it means 'to recognize Christ in the virtual incarnations of others as we make ourselves known.' Ott also helpfully updates gender violence realities such as digital sexual harassment and intimate violence. If you want to introduce Christian digital ethics with language for body-affirming sexuality and frank and accurate information, teach this book!"

— Traci C. West

author of *Solidarity and Defiant Spirituality*

"The digital revolution has added another dimension to the already complex area of human sexuality. Kate Ott provides a well-informed survey of the central issues at stake that will immensely improve readers' digital literacy. Offering a values-based framework, Ott has managed to avoid a *laissez faire* approach by which anything goes, while at the same time not becoming prescriptive. A first of its kind, *Sex, Tech, and Faith* is a valuable contribution to discussions on sexuality in the digital age that will assist Christians to navigate this aspect of modern life."

— Jonas Kurlberg

Centre for Digital Theology, Durham University

"Kate Ott, a leading Christian sexual ethicist and educator of our time, makes sexuality in the digital age—a prevalent hidden curriculum of Christianity—an explicit subject with which contemporary Christians must engage. In this accessible and profound book written for both adults and youth, Ott successfully persuades Christians of the need to be fluent in this critical but often-avoided topic to live an ethical and faithful life with powerful, bold, and realistic stories and challenging questions. Beyond the precious content, the beauty of this book is Ott's showcasing of embodied learning in her writing style through a multifaceted social-justice lens, a pedagogical principle that she advocates throughout the book. Theological teachers, students, pastors, and lay Christians alike will be equipped with both the what and the how of sex, tech, and Christian values—a rare gift readers will gain from this book that is both academic and practical."

— Boyung Lee
Iliff School of Theology

"With frankness and nuance, Kate Ott investigates the potentials of digital technologies for exploring and expanding sexual pleasure, identity, and care. Unabashedly sex- and tech-positive, Ott is at the same time highly sensitive to the complicity of the same technologies in racial, ableist, and gender-based injustice and violence. As she invites the reader to ponder novel and often-scandalized phenomena like sex robots, online matchmaking, cyberstalking, and digital pornography, Ott subtly recasts age-old questions about human relationality, vulnerability, curiosity, and embodiment. The double pursuit of digital literacy and a values-based theological sexual ethic nudges readers to cultivate their own erotic attunement through and beyond particular technologies. The shame-free yet deeply reflective tone, the sexual storytelling and cultural illustrations, and the accompanying discussion and reading guides make the book eminently accessible and implementable. The youth study guide in particular will be an invaluable resource for progressive faith communities."

— Hanna Reichel
Princeton Theological Seminary

"Kate Ott's *Sex, Tech, and Faith: Ethics for a Digital Age* offers clear instruction providing readers with the tools they need to cultivate their own thoughtful and faithful approach to sexuality in this digital age. Each chapter presents case studies and discussion questions to ease into sexual topics that are seldom discussed in Christian communities. Her frank delivery, her compassionate concern for bodies, minds, and souls, and her skill with practical advice make this book a must-have for anyone daring to learn about creating healthy sexual relationships."

— Monique Moultrie
author of *Passionate and Pious:
Religious Media and Black Women's Sexuality*

Sex, Tech, and Faith

Ethics for a Digital Age

KATE OTT

WILLIAM B. EERDMANS PUBLISHING COMPANY

GRAND RAPIDS, MICHIGAN

Wm. B. Eerdmans Publishing Co.
4035 Park East Court SE, Grand Rapids, Michigan 49546
www.eerdmans.com

Published 2022
Printed in the United States of America

28 27 26 25 24 23 22   1 2 3 4 5 6 7

ISBN 978-0-8028-7846-5

Library of Congress Cataloging-in-Publication Data

A catalog record for this book is available from the Library of Congress.

Contents

Foreword

I N HER NEW BOOK on digital sexual ethics, Kate Ott asserts the following:

> Sexuality includes spiritual, emotional, mental, and tactile experiences. This is what I mean when I say sexuality is embodied. The teaching of *imago dei*—humans created in God's image—forms the theological understanding of what it means to be human. When we speak of the *imago dei* in one another, it is a theological connection to or explanation of our embodiment. It doesn't reduce us to flesh, blood, and bones that are an exact replica of God. The *imago dei* is knowing and being known by God, evidencing God's presence in the world through embodiment—the tangled, integrated, and complex mess of spiritual, cognitive, emotional, and physical existence. (25)

With the emergence of a wide variety of digital and media technologies, the landscape of intimacy is rapidly shifting. It can be overwhelming. It can be exciting. For the church, it can often produce anxiety. And yet, what feels sadly lagging are the necessary fruitful conversations on how to engage these questions about what it looks like to navigate these changes in honest and compassionate ways.

Thankfully, in *Sex, Tech, Faith* Kate Ott gives us an incredible resource to question, to reflect, and to participate in these conversations. She does this by including topics that are often on the periphery of conversations about sex because of their taboo nature—topics like digital pornography, VR, and even sex robots—and by offering wise guidance on issues that have now become ordinary and everyday, like online dating apps. These are more than case studies—they are phenomena that are becoming deeply embedded in our culture and are shaping our identities and relationships. Ott's courageous approach pushes us to think *ethically* in a way that creatively considers the theological possibilities of the *imago dei*, embodiment and incarnation, creation, and ultimately questions about humanity in fresh ways. It is truly exciting. Her compassionate and thoughtful presentation not only shows us the complicated intersections of sexuality, technology, and theology by giving us relevant data; she also gives us language to use that is accessible and generative as she pulls insights from a wide range of sources.

If you are looking for a book with easy, straightforward answers, a black-and-white framework, or even simply a manual for behavior, this is not it. If instead you are like me and need a guide for asking new and different questions in faithful ways, *Sex, Tech, Faith* will be the book. At the end of each chapter she provides bullet points with strong, clear takeaways and gives us questions we can ask to cultivate our own thinking and reflection within our own contexts. At the end of the book is a wonderful and extensive bibliography as well as carefully prepared "youth studies" to be used in group settings with young people who are certainly navigating a similar landscape. Ott explains her goal for the book in this way: "This book's purpose is in part to inform the reader, but it is also *to invite them into considering their own Christian digital sexual ethic.* What values are core to you and why? Most importantly, how do you live out those values in a digital space? We will need to be open to moral creativity since we do not always know exactly how digital technology is changing us and our relationships" (12).

Most of all, Ott approaches this work graciously. With grace. There is not a whiff of judgment. But that does not mean that she is not rigorous or serious about the consequences of our actions and interactions. Her ethic is clear throughout the book—it is values-driven and praxis-oriented. It has to do with flesh-and-blood, with bodies and with our lives. She shows us that these conversations cannot happen only in an ivory tower or in a void. This is what makes the chapter on sexual violence, harassment, and abuse so powerful. She takes to task the theologies that undergird these structures and encourages us toward resistance work by showing us that it is possible to do this work faithfully and theologically and that we are called to pursue it.

Why do I trust Kate Ott? It is clear in the pages that follow that she is committed to work that furthers the collective endeavor to facilitate human dignity, loving relationships, and healthy communities. That she asks often "How do we make healthy, faithful decisions about our own bodies and how we view others?" is an indication to me that she cares deeply about the bodily well-being of people, and especially those who have experienced or currently are experiencing marginalization or violence by systems that continue to push singular narratives for how to be, how to love, how to live. Individuals and faith communities will find here a safe space to explore, and to imagine, and to embody the flourishing of our most intimate relationships.

Mihee Kim-Kort

Acknowledgments

W RITING IS A SOLITARY ENDEAVOR. And yet creating a book takes a community. Doing so during a pandemic requires support and fortitude. I have wanted to write this book for many years. As a Christian ethicist and sexuality educator, I started receiving questions at church gatherings when talking about sexuality and faith about how digital technology affects people's sexual and romantic relationships. Before I could answer these questions, I had to develop digital literacy. I postponed writing this book so I could pen *Christian Ethics for a Digital Society*. That book is a shared journey, with a style much like this one, to bring Christian values into conversation with how digital technology is designed, used, and adapted. Once I felt somewhat digitally literate—though rapid changes make it hard to keep up—I returned to the questions about how digital technology affects sexuality.

This project brings together two topics I love: digital ethics and sexuality education. The research for this book owes its foundation to theological insights from queer studies, womanist and feminist ethics, and Christian sexual ethics. Like a building's foundation, a significant amount of research hides below the surface, and I have translated it for the reader into everyday examples and faithful reflection. I invite the reader to join me in learning about topics I never thought I would include in a book. I'm not even sure sex robots were a thing when I was in graduate school, or at least not one I expected churches to have to address.

While sexuality and relationships rapidly change in response to digital technological possibilities, churches have for decades continued to fight over issues related to sexuality. This text does not participate in those debates. I invite the reader to discern their own sexual ethic while demonstrating how a values-based approach guides ethical expansion and moral creativity in response to digital technologies. The book explores liberating possibilities and explicitly critiques continued threats of heterosexism and gender binaries that often intertwine with issues of racism and ableism in sexual relationships and social structures.

Christianity is a rich resource for sexual health, wholeness, and morality. Unfortunately, many Christians turn away from church because they do not hear their church leaders providing relevant or inclusive sexuality information. Christian leaders may be uncomfortable and unprepared to have such conversations. This book is one starting point. And there are many additional resources throughout the text for further learning. The youth study guides are as useful for teen and young adult conversations as they are for adult education groups. The best way I know to share the good news about sexuality as a sacred, good, and pleasurable part of Christian ethical living is to talk about it. Break the silence, the taboos, and start talking. This approach is modeled in the text, addressing topics with frankness and care.

Every semester at Drew Theological School, I start classes by sharing words written by Christian ethicist Philip Wogaman. In *Christian Ethics: An Historical Introduction*, he ends the text by reminding the reader that "too much certainty about God's ways with humanity may not leave enough space for God to be God."[1] I try to approach every ethical question with this wisdom in mind. I hope the reader will do the same. Ethics is an art that requires creativity and flexibility, not a rigid rule book. The subjects of sexuality and digital technology constantly reinforce this truism, as what we know and experience related to both constantly changes, unfolds, and generates new insights and challenges.

I am thankful to Eerdmans Publishing, especially Andrew Knapp, for believing in the need for this book and having the vision to support it. I am grateful to have conversation partners who are the community out of which this book was created. I unexpectedly found a community of theologians in the Theologies of the Digital collective (https://cursor.pubpub.org) who value experimentation and collaborative feedback. That is a rare experience in academia, where competition and individuality are the norm. The group publishes work with histories of commenting and editing, while being fearless in its topic choices and theological discussions. It's where I try out some of the background theoretical approaches to the everyday ethics topics published in this book.

My appreciation for support and collaboration extends to many colleagues. Melanie Harris and Elías Ortega have sustained my writing spirit during the course of this pandemic, fielding my complaints, insecurities, and joys with grace and constant presence. I count myself very lucky to be at Drew Theological School. The faculty value cutting-edge research in theological education. Special thanks go to Traci West—whose scholarship is critical to my own work—for her ongoing encouragement. At rare moments, a faculty member gets to learn with and from other faculty members. That is my experience of team teaching with Mark Miller and Jennifer Quigley, the #GLST (Gospel Living and Social Transformation) team. I thank them for the countless meetings we started late because I was rambling about my recent research; their humor and curiosity kept me going. Not to mention, I have a few fun sermon ideas based on our conversations. Now I just have to get Mark to let me preach them in chapel! Last but not least, the institutional support of two fabulous student scholars made the research and production of this manuscript possible. I am indebted to Lauren Sawyer for her bibliographic and research contributions and Daniella Hobbs for her feedback on drafts and reference work.

I am abundantly blessed with the support of friends and family. My friend crew, while probably tired of hearing about sexuality education topics, is always willing to give me feedback and respond to my questions. It is not unusual for me to start a party conversation like, "How would you design your avatar in an erotic video game?" And then together we laugh, argue, and challenge stereotypes of race, sex, gender, and so on. Thank you to the Campchella crew for genuine interest in my work and feedback from "real people" about these topics. Last, my family provides an unwavering base of support without which I would not be able to write. They do extra chores and rearrange their schedules to honor my writing time. We spent countless hours together in quarantine during the COVID pandemic. I couldn't have picked a better group of humans with whom to cloister! They are funny, intelligent, and generous. It is our relationships and their extension into other networks that give me hope and sustain my faith.

Introduction

FROM SEXTING TO SEX ROBOTS, online dating apps to avatar meetups, digital innovation is rapidly changing sexual behaviors and relationships while Christian ethical responses lag behind. Digital technologies alter how we express and practice our sexuality, including the way we meet people, communicate with partners, and engage in sexual behaviors. Generally, churches have been all but silent responding to these major changes, with the exception of evangelical and conservative churches condemning online pornography use. In fact, progressive Christian faith communities are notoriously ill equipped to address healthy sexuality and relationships.

As Christians, whether we label ourselves progressive or conservative, we can do better! Many progressive faith communities think that if they take social action on behalf of LGBTQ inclusion and gender equity or implement sexual abuse prevention policies, they have adequately addressed sexuality concerns in their congregations. This is important work—very important work—but it is not enough. Advocacy and prevention policies do not teach Christians how to have healthy sexual relationships, how to apply their faith values to sexual decision-making, or how to understand and express their sexuality in faith-informed ways. While progressive Christian communities might criticize the dominant sexual ethic of conservative Christians as patriarchal or heterosexist, at the very least, conservative Christian communities are explicit about their beliefs related to sexuality issues and provide their members with retreats, education, and counseling on these issues. Unfortunately, the limited range of sexual experiences that are affirmed or allowed to openly flourish within a conservative Christian sexual ethic means many people are left out or damaged by these explicit teachings.

When it comes to digital sexual ethics, we need at least two things: to better understand sexuality and to build digital literacy. In order to gain that knowledge, we need sexuality education that explains the complex ways our bodies interact with other humans and technology. For example, online dating requires a different negotiation of body image than does meeting someone in person for the first time. Site platforms—whether Match .com, Tinder, or Bearwww.com—have design characteristics as well as community standards that mesh social and cultural ideals of beauty with technological choices about representation. A user's profile has a connection to the user's physical body that is constantly under negotiation. With this negotiation come new ways of embodying sexuality and using it to communicate with others. Clothes were once the primary technology; now digital options like avatars are infinite and confining at the same time. If online embodiment isn't confusing enough, Christians also face Christianity's ambivalent relationship to the body—sometimes repudiating it in a quest for spiritual attainment and other times valorizing it as God's good creation.

A Christian digital sexual ethic helps readers wrestle with pressing questions about the advancement of digital technologies, consider their effects on sexuality and relationships, and stay educated about sexual and digital changes. Sex tech is a huge industry valued at upward of $30 billion. Its impact on individuals and society cannot be ignored. From education to entertainment, there's an app to assist with better orgasms, breaking up with a partner, meeting the perfect match, or reporting sexual abuse; there's a device to check sperm health, have virtual sex with an avatar, or connect to a partner's vibrator in another country; or there's a robot to help with masturbation or serve as a companion. One book can't cover all sex tech. One Christian ethic can't either. Common rules-based ethical approaches cannot adequately respond to the rapid pace of digital technological innovation and growing awareness of sexual and gender diversities. Digital technologies reshape who and how we are in all aspects of life, including sexuality. They are not simply tools that we need to resist or control. Understanding the intersection of sexuality and digital technology requires a level of curiosity about sexuality, sexual behaviors, relationships, and digital technologies modeled through engagement with holistic sexuality education and digital literacies. It also requires a reconsideration of Christian sexual ethics.

Sexual Ethics

Many of us think of ethics as a science somewhat like mathematics. We plug the components of human actors and social circumstances into an ethical rule book, and voilà! We get a "right" answer for how to behave. For example, take a thirty-two-year-old Latino and a thirty-four-year-old Black woman on a first date at a local outdoor cafe. Add these facts into an ethical formula. Let's use a standard, dominant formula: abstinence from sexual intercourse until marriage. If so, the only output ethic we get is that they should not engage in intercourse. What else can they do; what should they do? What if the ethical formula is based on treating our bodies as temples of the Holy Spirit (1 Cor. 6:19–20) and avoiding harm to them (1 Cor. 3:16–17)? Perhaps they should enjoy a healthy, delicious meal for the sustenance of their bodies and then engage in multiple forms of consensual sexual behaviors—holding hands, kissing, caressing, mutual masturbation, and so on—celebrating the beauty and pleasure of each other's bodies as God's good creation and avoiding harm. Of course, neither of these formulas or ethics produces a singular way to be or behave. It tells us nothing of their commitments to other people, their past histories, their orientation, or their gender identity.

Now what if these are avatars and the users behind them are actually a twenty-eight-year-old white lesbian and a forty-one-year-old Black bisexual woman who, when dating men, prefers men of color because of the racism she has experienced from white men? What ethic do we apply here? Do the users have to be honest about the racial, gender, or age differences between their avatar and their offline identity? Are racial identities, sexual orientations, or age fixed categories that impact experiences online and offline? And if so, is this necessary information to determine whether a sexual encounter is ethical or not? Consider the difference if these two women are married to each other offline and they use online avatars to explore sexual fantasies, and if they know both the online and offline selves that participate. Based on the first ethic, does their marriage give them permission to engage in any sexual behaviors they choose? Does the second ethic prevent them from taking on bodies not their own based on how they dress or design an avatar?

Confused? It is difficult to think through all these factors and variations. The very possibility of such radical offline and online shifts raises numerous questions that sexual ethics, and Christian sexual ethics in particular, have not yet fully engaged. These very short scenarios magnify the shortcomings of treating sexual ethics like a math formula with a list of Christian teachings about right and wrong rules that one should apply in any situation. Christian digital sexual ethics need to be more flexible and creative.

The history of Christian sexual ethics is complex and dynamic even though some Christian leaders and denominations speak about sexuality as though Christians have always believed or taught the same thing. Christian teachings related to sexuality are influenced by scriptural interpretation, historical tradition, social context, personal experience, and scientific information. The emphasis or authority different denominations and traditions place on these sources produce very different teachings and doctrines. That said, we can still trace patterns related to Christian sexual ethics, such as how historical factors shaped the shift from celibate to married clergy during the Protestant Reformation or the religious justification of the dehumanization of African and indigenous bodies during colonialization. As Christians, we need to wrestle with the ways historical shifts shaped sexual ethics and consider the ways new changes influence today's Christian sexual ethics.

Christian thought generally has three starting points for creating a sexual ethic: behaviors, relationships, or values. All sexual ethics take into consideration these factors, but the starting point plays a significant role in shaping the overall ethic. For example, the oldest Christian sexual ethic is celibacy first. If one can't maintain that, then one can engage in sex for the sake of procreation. This is a behavior-based starting point. The first and best behavior is celibacy. The next best "good" sexual behavior is penile-vaginal intercourse for procreation. This ethic comes out of a variety of historical concerns in the early church, yet it is still dominant today. A behavior-based ethic says little about relationship qualities like respect, consent, or mutuality, and reduces a robust experience of sexuality to one behavior. It also has no helpful advice for the majority of people—children, youth, non-coupled people, and people who are infertile because of age or biological function. Even if one chooses abstinence or celibacy from sexual intercourse, there are still various aspects of sexuality that need to be discussed, including gender identity, sexual orientation, intimacy, and sensual expression of one's embodiment.

The starting point of relationships for a sexual ethic opens up the possibility of attending to a wider understanding of sexuality. Since the Reformation, relationships have taken on added significance and have often been a starting point for Christian ethics. Specifically, marriage and family have become more prominent topics in writings that dealt with sexual ethics. Concerns about the roles and duties of marriage partners to one another and to potential children are given prominence in sexual ethics. Inherently, this widens the scope of sexual ethics to more communal concerns and lessens the focus on singular behaviors. Unfortunately, a relationship starting point for a sexual ethic assumes all people are either in a relationship or moving toward one.

In the late 1900s into the 2000s, most Christian writing on sexual ethics embraced a robust understanding of sexuality and used a starting point of values. This approach affirms that we are all sexual beings from birth to death regardless of whether or not we engage in sexual behaviors. It relies on an understanding of sexuality as multidimensional and including at least five or more interrelated aspects.[1] *Intimacy*—the feeling of emotional closeness to ourselves, another, or God—and *sensuality*—the use of our senses to interpret and experience the world around us—are aspects of sexuality that begin developing from our first interactions with others. *Sexual orientation and gender identity* are both part of who we are and who we become over time. Numerous biological and social factors are entangled in how people understand their sexual orientation and how they identify and present themselves or are categorized based on the gender scripts of a given society. Another aspect of sexuality is *sexual and reproductive health*, which is unique to each person's experience, including various factors of development, access to care, absence or presence of disease, and so on. Of course, *engagement in sexual behaviors or abstinence from them* also shapes people's sexuality. That is to say, we share many of the same features of sexuality with others, and yet no two sexualities are exactly the same because no two people are exactly the same. Bodies, social experiences, and cultural surroundings further define people's sexuality. Sexuality is always an embodied experience. Here, I mean to stress that sexuality is intrinsically connected to material, fleshly, and ever-changing bodies. Unfortunately, the Christian tradition has been guilty historically of contributing to the idea that "detachment from bodies is often considered a sign of intellectual and spiritual maturity and a mark of true science and morality."[2] Similarly one might suspect that engagement in digital sexualities is an attempt to flee the body. A sort of ideal meeting of the traditional Christian goal to escape limited, sinful flesh while enjoying the pleasures of sexual desire in a virtual realm. On the contrary, digital sexualities are also embodied and rely heavily on tactile and neurological responses grounded in bodies. Some digital technologies enhance or diminish one aspect of embodied experience (like a focus on visual pleasure) and thus accentuate or downplay other aspects of sexuality. Thus, digital sexual ethics, like all sexual ethics, requires that we ask: How do sexual experiences affect all aspects of our sexuality including sensuality, intimacy, sexual and reproductive health, sexual behaviors, and gender identity and sexual orientation?

With a broader definition of sexuality, Christian sexual ethicists have started putting values first. A values starting point does not exclude people based on age, sexual orientation, gender identity, relationship status, or engagement (or nonengagement) in sexual behaviors—be they online or offline. Instead, ethical decisions about what is morally good or harmful are based on how one lives out specific values related to their relationships and behaviors. An example of some values that current sexual ethicists name are the free consent of partners, equality, commitment, fruitfulness, social justice, safety, mutual pleasure, and intimacy.[3] Others focus on specific Scriptures that provide values-based guidance on matters such as neighbor love and stewardship of God's creation, including our bodies.[4]

Let's test these starting points—behavior, relationship, and values. Most people would say domestic violence and marital rape are unethical. Is that true if we apply each of these ethical starting points? In a behavior-based sexual ethic where procreative sex is the ethical starting point, little is said about issues of domestic violence so long as sexual partners procreate. This ethic sees women's value as determined by their procreative potential, often setting up sexist assumptions that contribute to blaming women if they experience abuse. Even in a relationship-based ethic of heterosexual marriage, many US states and some countries protect marital rape because they reason that in a marriage a partner is agreeing to sexual behaviors into perpetuity. The current versions of these sexual ethics, which have either behavior or relationship starting points, have not prevented or even routinely condemned violence. Instead, a values-based approach explicitly condemns violence and abuse. Domestic and sexual violence is always unethical when consent, mutuality, and equality are requirements for moral goodness.

We do not need to test extreme—yet sadly too routine— examples of violence to prove a values-based approach is the most inclusive and helpful for sexual flourishing. Most people's experience of sexuality is one of failure (and sometimes harm) in relation to the ideal of a "good Christian adult" who is sexually satisfied, economically secure, heterosexually married with 2.5 children and a dog. The Christian tradition and congregational practice have for too long been used to prop up this ideal without providing helpful, faith-based education on sexual development or healthy relationships. And so, we fail, try again, fail, try again, and repeat. This might tell us two things. First, this particular ideal, based on one relationship and certain sexual behaviors, is the wrong thing for which to strive. Society, in the United States and beyond, is set up to allow only a few people to achieve this ideal. And high rates of divorce by those who "achieve" this ideal should signal that something is not working. Second, the dominant versions of Christian sexual ethics are unhelpful: celibacy or (heterosexual) marriage are not the only options.

What do we want instead? How about a strong relationship between ourselves and God that includes our sexual selves in all their fullness; feeling known and accepted in our communities; being empowered with knowledge about our sexuality; having the tools to communicate about sexual desire without embarrassment or shame, social and economic circumstances that support reproduction when chosen, freedom to form relationships that sustain us emotionally, spiritually, and sexually? And that's just the start. This sounds a lot better. Christian values like those named above can serve as guideposts throughout our lives to help us develop a healthy sense of sexuality and relationships with a fuller integration of faith.

The shared values named above are not unique to Christianity. One might argue that shared values like consent, commitment, and mutual pleasure should be universal or general principles that guide all sexual relationships. In this case, Christian ethicists claim these values come out of their Christian theological and scriptural interpretation, but they could also be sourced from other places like philosophy, human rights, or science and thus apply to people besides Christian believers. I trace these shared values in the Christian context back to core theological and ethics claims grounded in Scripture. Throughout the chapters, we will engage various theological questions about who we are as created beings and what God calls us to do and be in this digital world. Rather than picking and choosing various scriptural passages, I bring forth guiding scriptural themes and the traditions of interpretation and theologies that have arisen from them.

For example, we can consider the love commandment. Even this can be difficult to put into practice. Who counts as my neighbor and how do I learn to balance God, self, and neighbor? I have found that Cristina Traina's writing on erotic attunement provides a helpful approach to the difficult task of living out the love commandment in light of the robust, wholistic definition of sexuality provided above. She defines attunement as "perceptive attention and adjustment to feelings, needs, and desires—both one's own and others'."[5] Rather than temperance or control of the body or denial of desire, Traina suggests we should cultivate desires and better understand our bodies. Erotic attunement reclaims the pleasure of touch and does not renounce sexual desire and pleasure or suggest it is only for marriage or procreation. Erotic attunement is not a slippery slope to anything goes. Instead, attunement will mean there will be times when a person chooses to restrain their erotic desires as well. In this way, cultivating attunement leads one to be more attentive to power relations and justice demands. Traina notes that attunement requires a "constant motion between self-awareness and attentiveness to the other."[6] She looks to other feminist and womanist theologians who have reclaimed the erotic, suggesting that erotic love or eros does not objectify or use the other, the neighbor. Rather, erotic love "desires the person not as we fantasize her to be but as she is,"[7] neither seeking possession or control of her; and it "recognizes and pursues the beautiful as a kind of catalyst that brings forth the creative potential that already resides inside the lover."[8]

Importantly, attunement to erotic love requires practice as it "combines perception, imagination, and experimentation in an endless, partnered dance"—a dance that partakes in a self-correcting process.[9] Traina discusses attunement as something we strive toward, not something we achieve and never have to revisit. Traina suggests that intimate relationships evidence positive aspects and negative harms that are also present in the wider community. Thus, such relationships are microcosms of larger social issues, not distinct, privatized realms unaffected by social injustices. She writes, "If we really wish to reduce the irruption of violent, addictive, and depersonalized expressions of sexuality we must self-consciously cultivate just, relaxed, open, erotic sensuality."[10] If we can cultivate erotic attunement, we will be more aware of the needs and relational dynamics of self, God, and neighbor. Throughout this book, I will refer to ways erotic attunement shows up, needs to be cultivated, or is lacking in certain examples of digital sexual experiences, expressions, and relationships.

In addition, the cultivation of erotic attunement helps us recognize guiding values in action—values like honesty, mutuality, or commitment. That is to say, Christian values serve as guideposts against which one can check in on the moral formation of digital sexuality and relationships. They will not, however, provide a comprehensive ethic to apply in a "once and for all" manner. This book does not necessarily seek to say, for example, sexting is wrong because it violates honesty or does harm. Instead, an ethic of creative moral response seeks first to describe how we engage in the process of ethics as a form of creativity or play, and from that discern a praxis-based ethic of action and reflection in place of the rule-based approach. It asks: What is sexting? How do different technologies shape the practice? Why do people participate in it? How does it help or harm relationships? What does the practice suggest about a sacred view of bodies and creation? After answering these multifaceted questions based on a variety of user experiences, then we consider creative moral responses that guide human and digital interactions. Some will say this is too complicated or does not provide enough guidance. Often, that concern is grounded in a desire to control others' sexualities.

We must stop stifling the human capacity for moral creativity, recognition of the other, and desire for flourishing. Traditional Christian sexual ethics promote singular ways of being in the world that preserve power and privilege for a select few and that become meaningless with wider and richer experiences of sexuality and relationship. Likewise, digital technologies change so rapidly that ethical rules are quickly outdated. Moral creativity as an ethical approach means ethics is more like art or improv than mathematical rules.[11] This does not mean anything goes. Instead, I'm calling for a higher standard of sexual ethics with which one lives out values in difficult and complex social relationships with self, others, and God. In order to do that in a rapidly changing world, we need creativity and flexibility. I invite all of us to be Christian ethicists, thinking critically about how faith informs who we are and what we do as individuals and communities.

This book's purpose is in part to inform the reader, but it is also *to invite them to consider their own Christian digital sexual ethic.* What values are core to you and why? Most importantly, how do you live out those values in a digital space? We will need to be open to moral creativity since we do not always know exactly how digital technology is changing us and our relationships. This means I will take a praxis-based approach, which is an educational term for moving between moral action and reflection, to test moral values in different circumstances. Honestly, this is the process we already use in our everyday lives. We interpret what the world around us tells us about how things should be—dominant systems. Sometimes "what is" matches up with or differs from Christian practices and teachings. In addition to Christian teachings, we use other information from social sciences, personal experience, and so on to make decisions. Even a single person making a decision is informed by and often accountable to a wider community. Sexual ethics can feel like a test, and oftentimes we fail or fall short of an ideal outcome. That is the process of moral formation, and it happens on community and individual levels. And we can get better at it. My hope is that the approach of this book is familiar, affirming, and liberating all at the same time. That together we can clarify how to have whole and holy sexuality in digital contexts.

Between the Covers

Each chapter addresses a different digital sexuality issue by presenting current data on the range, use, and impact of the technologies, thus increasing the reader's sexual knowledge and digital literacy. Each chapter will also explore Christian sexual ethics literature paired with digital theological insights. The theological writings I choose often pay close attention to inequality and injustice. Gender and sexual inclusivity, as well as the fight to end sexual violence and harassment, will be core to the ethical assessment of digital sexualities for Christian flourishing. Each chapter will wrestle with concrete examples, some from research studies and others from digital fiction narratives, to spark creative moral responses.

These "stories" are shocking in some cases and include very frank talk about sexual behaviors and scenarios. Theologian Marcella Althaus-Reid suggests, "It is from human sexuality that theology starts to search and understand the sacred, and not vice versa."[12] She uses a method of sexual storytelling as part of what she describes as doing indecent theology. For some readers the stories I use will be shocking; for your average Netflix viewer they will be routine. I use these narratives for two reasons. First, sexuality is a taboo topic and stories can viscerally arouse hidden aspects of sexuality. Second, some of what we are discussing requires imagination because some technologies are still in development. Stories offer an encounter with the other, the neighbor, as the reader develops their sexual ethic.

As a Christian sexuality educator, I know many of us lack the ability to have frank and accurate conversations about sexual behaviors that also respect the sacredness of sexuality and relationships. This book strives to model how to do this. Where possible, I avoid euphemisms and slang unless their use helps the reader understand the experience of someone who uses those terms.

Also, I have a shame-free and pleasure-positive approach to all sexual behaviors discussed in this book. No behavior will be considered in and of itself as morally wrong; it will be assessed based on how it helps or harms self, others, and our relationship with God. There is a long history in the Christian tradition, for example, of condemning masturbation. Some people seeking scriptural support wrongly assess masturbation as the "sin of Onan," who was not masturbating but practicing the withdrawal method, or "pulling out." Rather, I attend to both scientific and theological research that considers masturbation a healthy and moral good leading to education, empowerment, and sexual pleasure. That does not mean that all instances of masturbation result in a health or moral good—context matters! We discuss this in more depth in chapter 1 in the context of using sexually explicit material.

Chapter 1 focuses in on a widely deliberated topic: professionally produced and self-made digital pornography. This topic intersects with issues of sexting, alternative sexuality sites used for education and erotica, body image and sexual stereotypes, consumption of sexual violence, and more. We explore how digital technology changes the human experience of consuming and producing sexual material, including ways online spaces contribute to and disrupt anti-Black, ableist, sexist, and heterosexist stereotypes. Bringing theological questions to this work, I center on what it means to be made in the image of God and how that relates to the incarnation. In response to the liberative possibilities and violent harms of professional and self-produced digital pornography, we can uncover Christian theologies and ethics that are body-affirming and reinterpret nakedness without shame in the Eden scene. In order to do this, I invite the reader to play with creative differences between lust and erotic affirmation for bodily createdness.

The next chapter focuses on the use of digital relationship sites and applications to find partners for offline sexual and romantic relationships. Hookup, dating, and marriage sites often employ digital technologies that delimit how users present themselves digitally and deploy complex yet stereotype-laden algorithms to generate "matches." The creators of these sites are invested in utilizing impression management, attention growth techniques, and rewards to generate relationships. I ask how these technologies shift communication and meaning-making in sexual relationships. Scripture often uses language like being known or belonging to another as metaphors for sexual and romantic unions. What forms of moral knowing happen online when we are embodied differently—or, some might argue, disembodied? Christians have for centuries had much to say about the need to get married and about who should be allowed to be married, but sadly less has been written about healthy sexual relationships of all sorts. Online relationship sites have allowed for innovative ways to meet people. We need to consider how these platforms change, strengthen, and complicate faith-informed sexual relationships.

Digital technologies that bring people together can be equally used to follow, surveil, and exploit others. This chapter looks at the violence perpetrated through the technologies discussed in other chapters. Examples include deepfake revenge porn, cheating applications, harassment in direct messages, and sexual violence in virtual reality or against sex robots. Each chapter briefly attends to technological vulnerabilities for exploitation, but this chapter goes into further depth about the nature of sexual violence, harassment, and abuse. It also invites readers to consider how Christian traditions of self-sacrifice and enduring violence can be wielded against victim-survivors of abuse and harassment. Moral imagination often falls short of conceiving of a world free of violence, falling back on the limitations of sin. However, the recognition of failure does not require us to promote or tolerate abuse and violence. It is long past time that faith communities promote spiritualities of resistance to sexual violence, harassment, and abuse online and offline.

In chapter 4, we consider the implication of embodied sexual behaviors and relationships in a digital space including virtual and augmented realities. This includes relationships or encounters that happen in an immersive online space, that involve forms of sexual behaviors utilizing avatars or virtual representations of the self, and that engage an embodied response from the user. Users react emotionally, physically, and psychologically in virtual spaces in ways that affect their offline sense of self. Might there be liberatory ways to use these spaces to promote social goods like decreasing sexually transmitted diseases and unintended pregnancies; allowing users to explore different genders or sexual orientations; or providing practice at relationships including sexual behaviors to gain facility and skill, enhancing offline connections? Scriptural traditions refer to sexual intercourse as "knowing someone," giving us a richer conception of the role emotions and bodies play in learning and connecting. Jesus's incarnation is an example of being known in a different embodied reality connected to God's divinity. Jesus is not only incarnated in his historical time period, but is also met through each of us as we encounter one another. What does it mean to recognize Christ in the virtual incarnations of others as we make ourselves known?

Last, in chapter 5 we move humans offline and discuss various embodied artificial intelligences such as sex or companion robots. It's often noted that these technologies have been created for altruistic purposes because everyone deserves the opportunity for love, companionship, and physical intimacy, which for a variety of reasons can be easier with an AI device than a human. For example, when intellectual or physical disabilities lead to discrimination and loneliness, should those with such disabilities be able to find sexual satisfaction and even love by turning to artificial intelligence robots? Does AI remedy this human-generated social issue or exacerbate it? Christianity has for centuries been more concerned with legislating sexual behaviors than addressing the affective needs of adults to find closeness and intimacy. Of course, when we look more closely at such technology, we quickly realize that gender, racial, and ability standards of beauty affect the design and purchase of AI sex robots—to say nothing of the need for user data collection to improve the AI. If we are co-creators with God, what are the limits to making another in our own image or in the image we prefer? Challenging the boundaries of moral imagination, we can infuse artificial intelligence with an ethical formation that reflects back to humans our own creative foibles and follies—similar to how humanity provides clues to the nature of God.

Each of these chapters could be turned into its own book covering either the technological issues or the theological considerations. While I will provide helpful background and current research on these topics, none will be exhaustive. That is why each chapter will be accompanied by discussion questions. And I include a suggested reading list for further investigation. As I have noted, the ethical approach in this text invites the reader to learn and grow through creative moral engagement, something we often forget to do or stop doing as we get older. As sexual beings from birth to death, we have plenty to keep exploring; we simply need to make it a priority and truly embrace all of our createdness at every life stage.

While youth and young adults are often the focus of sexuality education, they are also a generation coming of age with digital sexualities. In order to engage them in this conversation, there is a guide for using this book with young adults and youth—including a lesson plan that addresses the main theological issues, pulls out age-appropriate sexual and technological information, and provides discussion questions. It is available at the end of this book with references to further age-appropriate resources.

As Christians, most of us fumble around with the lights off(line), hindered from exploring the fullness of our sexuality. In response, more and more people seek online spaces to discover their sexuality and engage in new sexual practices. As a seminary professor, Sunday school teacher, and Christian sexuality educator, I want to better understand how digital technology has changed the way we initiate, engage, and sustain sexual relationships so that I can responsibly bring my faith values to these important, intimate connections in my life, in the lives of future and current faith leaders, and in congregations. If you have picked up this book, you share this commitment. Let's get started!

In the Image of God

TO CONSUME AND MAKE DIGITAL PORNOGRAPHY

TECHNOLOGY IS OFTEN THOUGHT OF as a tool that humans manipulate and control. Rapid proliferation of technological devices, access to them, and programs for them have, however, reinforced what humans have known for years: technology shapes us as much as we shape it. Have you ever swiped the screen of a laptop, forgetting that it isn't a touch screen like your phone's? Do you expect an instant response when you send a text? Have you lurked around Facebook checking in on old friends, wondering what their lives are like now? Social media and mobile technology have changed our behaviors and expectations, as well as our standards of privacy.

Related to sexuality, no other aspect has been more revolutionized by digital technology than pornography. Pornography has been a part of societies for thousands of years, from sculptures and wall drawings to print and magazines to moving pictures and DVDs. Each new technological innovation more intimately portrays human bodies and sexual behaviors and increases the privacy of the consumer. Digital technology exponentially exacerbates these two features—imagery and privacy. All forms of pornography are available on multiple digital platforms that can be accessed without others being aware of the viewer's behavior.

Overexposure to sexualized images through everyday media makes it more and more difficult to know the difference between sex-affirming or sex-positive versus exploitative material. Without a counterbalance to the ubiquitous access to and stereotypical portrayals in media, people can adopt a commercialized and objectified approach to bodies, sexual behaviors, and relationships. The church's two primary responses—silence or blanket criticism—limit discussion about body-positive theologies and cultivate shame. Additionally, many adults and youth are (knowingly or unknowingly) creating pornography via selfies, sexting, and video media such as Snapchat. Self-made pornography is not new, but digital technologies make it easier to produce and disseminate.

How do we make healthy, faithful decisions about our own bodies and the way we view others? Objectifying another person's body or sexual acts dehumanizes and degrades the sacredness of sexuality. Some pornography eroticizes violence and features degrading behavior (often toward women), exploitation, and subjugation for the purpose of sexual arousal. These features are what make some forms of pornography illegal. Yet not all sexual material fits this description. There is a difference between material depicting mutual erotic activity and material that is violent, sexist, or racist. Some sexuality experts distinguish pornography from erotica—a category of sexual material that does not harm or objectify persons or sexual behaviors. Erotic materials like these are often a teaching tool used in sexuality education and counseling. A primary way to find out more about bodies and sexuality is to observe other people. In healthy and loving relationships, use of erotic materials may enhance a couple's sexual relationship. Thus sexual material is diverse, and it should not all be de facto judged illegal or immoral.

Does viewing a person naked or watching someone perform sexual behaviors turn them into a sexual object? There is a perception that the Christian answer to this question is always "yes." Many cite Jesus's response in Matthew 5:28: "But I say to you that everyone who looks at a woman [or anyone] with lust has already committed adultery with her in his heart." However, what's missing is a nuanced understanding of lust. When we look at another person, we are called as Christians to see the image of God in them and to value their beauty in physical, emotional, and spiritual ways. Finding another person attractive is not an automatic objectification of them so long as we are valuing them as a whole person and affirming their beauty as a reflection of God's creation. Similarly, enjoying sexual pleasure is not implicitly exploitative or sinful so long as we are not engaging in that behavior for personal gratification *over and against* another person's pleasure and consent. Even self-pleasure or masturbation can be empowering, as it helps us understand how our bodies work and perhaps be more open to communicating with a partner about sexual pleasure, so that pleasure will be mutually experienced. In this chapter, we will explore sexual theologies and ethics that provide the foundation for a body-affirming sexuality.

Even if one seeks to avoid all sexually explicit material, sometimes it finds you, through pop-up ads, forwarded posts, or email spam. More often than not, adults seek it out. Pornography use may not be talked about in faith communities because of sexual taboos or lack of education, but those who fill the pews are definitely accessing it. And it's not just adults. Research published in *Pediatrics* found that four in ten tweens and teens visited a sexually explicit site over the course of a year. While boys were more likely than girls to seek out pornography, two-thirds of the time exposure to images was unwanted.[1] Findings suggest that use of online pornography by youth may skew their understanding of bodies, sexual behaviors, and relationships.[2] Similar findings have been reported for adults and will be discussed in more detail later in this chapter.

The use of digital technology to view pornography raises specific issues. Online pornography can result in legal infractions either because of the age of the user, the age of the actors, the type of content or the device used to view it. For example, using a work laptop or a school computer (though most block sexually explicit material) can result in criminal charges regardless of the legality of the content. In many cases, it is the specific content that determines legal punishment, and that most often relates to child pornography. This is an important issue. However, this chapter will not focus on *explicitly illegal* pornographic material such as that which features children. Ease of production and distribution leads to difficulty tracking down those who produce child pornography or participate in sex trafficking, which is something I address in chapter three. That is not the predominant use of pornography, and there is overwhelming consensus in faith communities on its immorality. While this chapter focuses mostly on the user of pornography, digital technology has caused rapid changes in the pornography industry, creating a gig economy that gives sex workers more autonomy.[3] The creation of pornography now includes diversity in labor, production, and distribution of materials—from traditionally produced videos or photos where workers have less control over the product to video games, members-only websites, virtual reality, webcamming, and Only-Fans, where the worker, or the "star," controls production.

Sexually explicit material online is extremely diverse. The focus of this chapter is whether or not *legal* sexually explicit material can have an educational purpose or moral value in ways that witness to the image of God in other human beings. First, online platforms make a variety of pornography widely accessible—able to be viewed in private and affordable (even free)—increasing its use among all segments of the population, some benefiting and others experiencing negative outcomes. Second, research suggests that the speed at which online pornography can be viewed contributes to compulsive behavior for a small, specific demographic of users, impacting their future sexual satisfaction with self and partners.[4] This chapter will delve deeper into the science of this phenomenon and social responses. Third, innovations in technology have also brought about the "personalized pornography" boom. Sexting is the popular term for sharing sexual content such as words or pictures, or for live interaction via social media. This new form of sexual communication can have unintended adverse consequences in some contexts and can promote relational well-being in others. We need to have conversations about how or whether various forms of pornography enhance or deny values of honesty, consent, and mutuality in relationships, as well as how they impact our self-image and expectations of sexual performance. Even though creating, viewing, and sharing sexually explicit material is an age-old practice, digital technology shifts current experiences.

Affirming the Body and the Image of God

Sexuality is about the whole person, something even our Christian traditions have often neglected. When we do not openly talk about the connection between sexuality and spirituality in faith communities, we perpetuate a shame-filled and dualistic approach to sexuality and bodies. That affects how everyday Christians live their sexualities. Consider the following typical young adult experience.

> Kicking off her shoes and removing her coat, Danielle collapses on her bed, exhausted and bored from another long day as an office manager. She rubs her brow, tired of solving everyone else's problems during the day, and sinks into mindless scrolling. She opens Instagram. Her thumb flips photos upward. Then a photo piques her attraction. She switches over

to a free porn app and searches using the same features that piqued her interest in the IG photo. She begins to masturbate to the fourth or fifth option, dismissing the other videos after only watching a second or two. She knows her friends do this too, but they don't really talk about it. Easy access to free porn on the internet sometimes means Danielle has to scroll a while to find something she likes. She doesn't mind because it's private on her phone. Danielle is a thirtysomething who goes to church periodically but was much more active as a child and teen in Sunday school and youth group. She doesn't see her faith as having anything to do with sexuality, but she does think the stuff she learned about girls being temptresses makes her shy about her body. She's never felt comfortable sending a nude photo to a partner even though she does when asked. She feels guilty about the fact that she hasn't married yet. As a biracial, straight woman, Danielle uses porn primarily for self-pleasure, which helps her reduce her stress and chill, without all the hookup drama. She feels too shy to communicate what she wants sexually in a relationship. So she started using porn to explore how her body responds when she masturbates. She avoids the stereotypical porn types because she feels affirmed when seeing different racial identities and body types, like those featuring big beautiful women (BBW).[5]

Even though religious experience is often sensual and embodied, most Christians struggle with a legacy of spirit-versus-body dualism. Danielle's statement that she doesn't see her faith having anything to do with her sexuality isn't an anomaly—she hasn't experienced a faith-based message that connects with her experience of sexuality.

When we shift our focus to Jesus, God incarnate, we can see the necessity of bodies for the Christian life. It is not in spite of bodies, but because of them that we can love, birth, pass peace, anoint, nourish, and relate. Pastoral theologian Bonnie Miller-McLemore writes, "Adults are not all that different from children, although we like to think we are. We do not leave sensate experience and knowing behind, even though Western doctrinal and intellectual history implies that such detachment is possible and even admirable. Our own religious convictions are more deeply buried in our bodies and bodily practice than we realize."[6] Far from denying or ruling over the body, sexual theology and ethics needs to promote a deepening understanding of how we live in bodies and relate to other bodies.

Sexuality includes spiritual, emotional, mental, and tactile experiences. This is what I mean when I say sexuality is embodied. The teaching of *imago dei*—humans created in God's image—forms the theological understanding of what it means to be human. When we speak of the *imago dei* in one another, it is a theological connection to or explanation of our embodiment. It doesn't reduce us to flesh, blood, and bones that are an exact replica of God. The *imago dei* is knowing and being known by God, evidencing God's presence in the world through embodiment—the tangled, integrated, and complex mess of spiritual, cognitive, emotional, and physical existence.

Embodiedness has become shorthand for the material ways humans are known or marked by sex, gender expression, race, and disability. Often society and religious communities use these external identifiers to create policies and practices related to who is included or excluded in communities. Like Danielle, we often look for spaces where we see other bodies like ours reflected as beautiful, sensual, and desired. The majority of pornography tends to valorize certain forms of white, able-bodied, young, physically fit, and sexually well-endowed bodies. In these cases, we must peel away the stereotypes of beauty that determine people's worth and further diminish the intended diversity of God's creation.

Embodiment is far more than these social categories.[7] That is to say, we don't just have bodies; we are bodies. And we often use technologies to extend the limits of our bodies. When interacting with technologies, we are not always cognizant of how we are being reshaped in the process. For example, when a person uses a shovel to dig a hole, this is a tool under their bodily control. And yet, they become a much more efficient hole digger when using a shovel and probably later in the day, their hand is aching from the reshaping of muscles and the blisters. Their identity and physical body are changed in the process. These types of embodied changes are exponential and more hidden with digital technology. For example, will Danielle's use of digital pornography enhance her self-image, and will it lead to greater sexual pleasure when she is in-person with a partner? We don't know the answer to that, yet; research shows it is different based on types of consumption, but pornography use does have an effect.

In "A Theology of the Body for a Pornographic Age," theologians Rhea and Langer note, "From Plato to Augustine to the Council of Chalcedon, the church and the general intellectual culture from which it has grown have found it difficult to grasp the ways that the physical body relates to the spiritual dimension of human life."[8] This is why past sexual ethics have focused on directing sexual desires toward one behavior or practice such as procreation, celibacy, or temperance. Sexual desires were originally thought to be part of humans' sinful nature and a distraction from relationship with God. In both Paul's first letter to the Corinthians and Augustine of Hippo's influential writings, we hear about sexual desires as overwhelming and the need to choose celibacy or confine sexual behaviors to marriage. We can find similar teachings in today's teen sexuality curriculum, like those Danielle experienced in youth group.[9] For ethicist Christine Gudorf, the ongoing perpetuation of teachings about sexual desire as a slippery slope of sinfulness and the body as something to be ruled over and controlled lead to the taboo nature of sexuality, and "The silence and shame—breeds ignorance."[10] Continuing the silence produces more shame and hinders one's ability to fully participate in sexuality as a good, created by God.

To honor the fullness of our embodiment and the expansiveness of sexuality beyond only sexual behaviors, we need something more akin to erotic attunement discussed in the introduction. Attunement requires critical engagement with how our sexual desire is elicited and engaged through sexually explicit material. Are we using it to better understand our own desires, needs, and bodily responses? Or do we consume it without reflection and awareness of the embodiment of those producing it? Danielle seems to have fumbled onto this insight on her own as she turned to pornography and masturbation to better understand her own sexual desire and pleasurable embodied response. Erotic attunement reclaims the pleasure of touch and does not seek to confine sexual desire and pleasure or suggest it is only for marriage or procreation.[11] The connection between personal use and larger social conditions is extremely important to recognize. In Danielle's circumstance, she knows the difference between fake model bodies and a more normal range of body types, and she instead searches out examples that show different body types that are more representative of her experience.

Achieving erotic attunement is something most of us would struggle with in an offline context. It requires we tap into a suppressed theological tradition that values embodiment, trusts human ability to cultivate desire in meaningful and fulfilling ways, and engages a multisensory and tactile experience of knowing. This is difficult work for those who suffer from Christian shame and taboo about sexuality mixed with a severe lack of quality sexuality education. No wonder Christians are confused and reduce sexual ethics to a few quick, blunt rules. In response, people in the pews go looking for answers elsewhere. As with Danielle, one of the primary places adults and teens go for sexuality information is pornography websites.

Online Pornography

Danielle probably isn't the stereotype you imagined of an online pornography user. According to research on women's use of pornography, however, she is fairly typical. I of course added the experiences of sexuality education in a faith context. Joel and Peyton are also typical users of pornography:

> Joel started using his computer for porn when he was a teenager. Mostly, he wanted to be like all the other guys who had seen it and knew what sex was about. He was also trying to figure out his own sexuality. By his mid-twenties, he had had a few girlfriends. He would sext with them, thinking it was common behavior to show your interest in a girl and gauge her commitment to you. But he never really felt sexually fulfilled with them. When he watched porn, he could easily orgasm. He started to get worried something was wrong. Now in his early thirties, Joel gets pressure from his Latino family about getting married and having babies. He decides to talk to his current girlfriend about watching porn together as a way to spice up their sex. His girlfriend is skeptical about the idea. So she asks Joel what attracts him to porn, because she worries it will make her feel objectified and unattractive. He goes on and on about the men's behavior. In response, she caringly and sincerely says, "I'm OK with trying it—if you will honestly ask yourself how sexually attracted you are to men."

> ———

> Peyton's not a huge fan of online porn sites, but they have nowhere else to go to see what a trans person looks like having sex. They've never even been in a relationship. Everyone at their school still thinks they are a "she" except for one close friend and the campus minister. Peyton's campus minister is progressive and supportive; Rev. Jessica has brought in speakers and started an LGBTQ+ (lesbian, gay, bisexual, transgender,

queer, and more) support group. Yet the focus is all about how to come out to your parents and friends and how to have a relationship with God even when your home church has kicked you out. Important things! What Peyton really wants to know, though, is whether a strap-on will make them feel like the person they are inside when having sex with someone they love.

These are "normal" uses of online sexually explicit material. That is, as much as we can talk about normal pornography use. It might be better to talk about trends and stereotypes to unpack what's really going on with people's use of sexually explicit online materials. The experiences of Joel and Peyton reveal additional dimensions of why and how individuals and couples use pornography. While people intentionally engage pornography for specific purposes, technology has always been "instrumental in how we experience, mediate and make sense of our sexualities," whether that be sculpture, photos, printing, books, video, or virtual reality.[12] For some, all sexually explicit materials are deemed as objectifying and sinful. For others, there is a category of sexually explicit materials, such as the fully naked and anatomically correct David statue, that express or appreciate the beauty of God's creation.

Digitizing sexually explicit materials, however, not only changes the type and scale of materials produced; it also changes how humans interact with the material that leans toward consumption more than appreciation. Access to digital pornography is affordable (often free), readily accessible through private devices, and scaled beyond demand. That is to say, all types and a seemingly unending supply of digital images and videos are available at one's fingertips. This allows Peyton to see themselves reflected in trans experiences and bodies at the same time as Danielle scrolls past violent, sexist, and racist portrayals of women. For some, the speed and privacy of access can cause obsessive viewing. For example, on some sites built-in software displays the next image or video in an automatic queue, and the next video starts automatically unless you close out. These attentional behavioral software features are designed to affect users' brain chemistry, and they do. Like Joel, the majority of pornography consumers are male-identified. Yet people across genders, sexual orientations, races, ages, geographic locations, classes, educational levels, physical and mental abilities, and religions use pornography. These diverse factors influence how and why people use sexually explicit materials which in turn impacts how it shapes them.

Pornography use is difficult to compare across populations and time periods. A few factors taken together can give us a general impression of its ubiquity. First, the self-proclaimed largest pornography website, Pornhub, ranks in the top ten most trafficked websites along with the likes of Google, Amazon, Facebook, and Wikipedia.[13] Second, composite research of national studies in the United States reveal that "46% of men and 16% of women between the ages of 18 and 39 intentionally viewed pornography in a given week. These numbers are notably higher than most previous population estimates."[14] These estimates relate to weekly use rather than if one has ever been exposed or periodically uses pornography; those rates are much higher. Third, pornography use is less culturally taboo than in the past, but still very private. When asked about the purpose of pornography, a survey by PBS/Frontline and the Kinsey Institute for Research in Sex, Gender and Reproduction at Indiana University found "86 percent of respondents said porn can educate people, and 72 percent said it provides a harmless outlet for fantasies."[15]

While most people use sexually explicit online materials in private, there are many who share pornography use in their relationship. The majority of research on this subject focuses on heterosexual couples. There are different results based on the gender of the viewer. That is to say, male use was sometimes negatively associated with sexual quality for both men and women, while female use was positively associated with female sexual quality.[16]

For some women in heterosexual relationships, their male partner's use of pornography can relate to lower self-esteem and body
image, which affects relationship quality and sexual satisfaction.[17] We can hear these concerns in the initial response of Joel's
girlfriend when asked if she wants to jointly watch pornography.
After hearing that Joel's desire to watch pornography might relate
to a wider understanding of his sexual attractions, his girlfriend
agrees, with the caveat that he needs to explore this. In other
studies of younger adult populations, there is a general positive
self-perception of the use of pornography and relationship satisfaction. Some data shows small adverse effects like decreased
relationship satisfaction for those with larger body types, those
needing longer to get stimulated, or those needing more sexual
stimuli to reach orgasm.[18]

Women and queer-identified users of sexually explicit online
material are increasing.[19] A recent review of studies on women's
use of pornography shows the main reasons women use online
sexually explicit material includes "fulfilling sexual needs; em-
bracing and exploring sexual selves; connecting in sexual relationships; and normalizing sexual desires."[20] Narrative engagement of a small representative sample of diverse women shows
that pornography use informs their sexual relationships, increasing communication about desires, new ideas, and limit-setting.
This is exactly the type of outcome that Danielle hopes for in her
use of pornography. Because of sexual taboos, women report that
privacy of digital devices is key. They may experience privacy
when viewing materials, yet most free sites track users' data so
every user account generates a record based on their digital data
trail. Female users on free sites also sift through a variety of materials to find what they like rather than relying on curated sites
that might cater to their values or desires, like feminist sites that
exclude violence, racist themes, or sexist behavior.[21]

LGBTQ+ populations have even less research published about
their use. Review of available studies suggest that Peyton's experi-

ence is both common for the desire to see themselves represented
and gain sexuality education. Unlike Peyton, other LGBTQ+ youth
find pornography earlier than their heterosexual and cisgender
counterparts and use it to explore sexual pleasure. Interestingly,
these patterns are not associated with similar negative outcomes
that affect heterosexual youth—such as pornography addiction,
poor body-image, and sexual dysfunction.[22] In fact, earlier and
more frequent use by LGBTQ+ populations has more to do with
resisting a heterosexual and gender-binary culture that stifles
discussions of gender and sexual diversities. Use of pornography
may contribute positively to LGBTQ+ individuals' sexual identity
development, so long as diverse pornography is available such
that negative stereotypes are not reinforced by viewing predominantly heterosexual materials.[23]

Evidence suggests the use of sexually explicit online materials
is often for education and sexual self-exploration, as well as related
to stress relief and relationship development for adults. These uses
are amplified by the benefits of digital technology—affordability,
viewing privacy, and diversity of media, all contributing to greater
accessibility. In the face of inadequate sexuality education, these
uses seem to contribute to greater sexual flourishing, including
connection with one's body and positive experiences of sexual pleasure. There may also be negative consequences like poor body image or periodic sexual dysfunction when one's experience is skewed
by online standards producing unrealistic offline expectations.
That said, it's difficult to discern how to weigh the benefits versus
possible downsides when negative body image and performance
standards are affected by other media images beyond pornography.
Sexualization and unrealistic sexual and relationship standards
are ubiquitous in mainstream media. To combat that, some forms
of pornography affirm diverse bodies and sexualities or empower
users to affirm their own sexuality and embodiment.

Nevertheless, the research and accompanying vignettes, created as composites to represent the data, suggest it is men, usu-

ally heterosexual, and to some extent women in relationship with them, who uncritically engage pornography and experience negative sexual outcomes. The research on "pornography addiction," a still-controversial diagnosis in the health care field, supports this observation. A few factors related to digital technology—a user's particular personal attributes and types of exposure—come together to make a compelling case for understanding some forms of pornography use as compulsive. This is a minority of users. In other words, Joel, who used pornography for education and exploring his sexual desires, experienced a lower level of sexual satisfaction in his offline relationship, but he did not experience addiction. This could be because sexually explicit online materials allowed him to explore bisexual attractions that were not present in his offline heterosexual relationship. However, a user like Calum, the "sex addict" in the British documentary film *Porn on the Brain*, turned to pornography to boost his self-esteem and vicariously live out sexual fantasies of dominating women with whom he could not successfully relate in person.[24]

Those who experience compulsive pornography use often bring some common factors to the experience. The majority are heterosexual men who began using online pornography at a younger than average age.[25] They have a skewed perception of body image for both men and women in offline contexts, and their pornography use is either reacting to or causing sexual communication issues in a current or past relationship.[26] Additionally, and important for the discussion of sexuality and faith, those with religious backgrounds who feel shame or moral judgement toward pornography use are more likely to assume they are experiencing "addiction" to pornography even when their patterns of use do not support this self-assessment.[27] Heterosexual men who report "problematic pornography use" tend to perpetrate physical and sexual violence, including intimate partner violence, at higher rates.[28] This data reflects a corollary, not causation. Research also suggests that some users are predisposed to overstimulation and addiction-like reactions to attentional behavior technology (like the blinking light or small image that signals you have a new message, or the constant cycling of new videos).[29]

In response to growing research that online pornography use is correlated with sexual dysfunction and violence for mostly male, heterosexual users, the World Health Organization added compulsive sexual behavior disorder (CSBD) to the International Classification of Diseases in 2018. Categorized as an impulse control behavior rather than addiction, CSBD "includes a persistent pattern of failure to control intense, repetitive sexual impulses or urges resulting in repetitive sexual behavior.[30] CSBD relates to online pornography use as well as offline behaviors. This classification helpfully allows for a diagnosis that can lead to care, even if the exact cause and effect relationship between digital pornography use and biological and social factors is not yet known.

To conclude, online pornography use can have positive and negative effects on our sexual embodiment. For the majority of users, the digital qualities of online pornography—like accessibility, affordability, diversity, and privacy—enhance the positive embodied impacts like reducing stress, cultivating sexual arousal, and affirming the sexual identity of a user related to body type and gender identity. For a minority of users, digital pornography use has negative physical effects, including brain chemistry changes, negative body image, and sexual arousal issues such as erectile dysfunction and delayed orgasm. While we must be conscious of the outsized impact digital pornography use has on a particular group of users, this should not lead us to condemn all pornography use. Rather, we need to talk more openly about the harms of the kind of pornography use that reduces sexuality to physical sexual stimulation or objectification of other human beings. We can affirm digital pornography use that presents a diversity of body types, orientations, and gender identities. Key features of ethical engagement with digital pornography are intentional connection or appreciation between the user and those depicted, insistence on holding relational concerns with self and others central to the experience, and development of a richer imaginative fantasy life. These lead to greater sexual flourishing for the user and their relationships.

Sexting

Some researchers suggest that sexting—sending sexually explicit written or visual content via digital media including email, direct message, or social media platforms—has become more popular because of the boom in online pornography. Perhaps increased access and exposure to sexually explicit online materials increase the prevalence of sexting, but what's more likely is that digital devices are our primary form of communication, including in sexual relationships. The capacity of digital devices to communicate by word, emoji, visual image, or video creates the opportunity for sexting to become a new form of sexual communication. More and more people rely on digital forms of communication to stay connected. This is especially true for adolescents, who have less freedom of movement to meet up with friends or potential partners. Though the overlap between sexting and pornography may not be causal, they certainly impact each other, as does the variety of sexual content across mass media formats. Perfume and clothing ads, Instagram profiles, and personal sexting photos all contain examples of similar sexually seductive poses.[31]

Sexting is considered a form of self-made online pornography. There are several other forms of self-made online pornography, including amateur videos of groups, partners, or individuals engaging in sexual behaviors, as well as sexually explicit images. This sexually explicit material is created specifically for wider public dissemination, whereas sexting is an interpersonal communication. A significant amount of research and debate has considered adolescent and young adult sexting practices, but much less attention is given to adults' sexting. This is most likely a result of the legal and sometimes social conflation of adolescent sexting with child pornography[32] and the significant negative social outcomes associated with secondary sexting—the intentional resharing or reposting of a sext to a wider group. Research also suggests sexting contributes to gender inequalities in heterosexual sexual encounters, harming young heterosexual girls more than boys. However, this may be an overly simplistic interpretation.

As with arguments about pornography and ethics, some will argue that any sexually explicit material is immoral and thus any form of sexting is wrong. This line of argument reinforces shame related to bodies and sexuality more generally. Instead, we begin from the point of view that our bodies are created by God and not only good but beautiful. They are our primary means of interacting with one another and sharing our attraction. With this starting point, it's no wonder that sexting has become a normalized sexual behavior for adults who primarily use digital devices to communicate with one another. Both Joel and Danielle have sexted with partners. Of course, concerns exist, such as privacy of the communication or future use of the material when a relationship ends. In an ethical evaluation, we should also be concerned with issues of power. Simply because the sender and receiver are adults does not mean the sexting is consensual. Danielle reports not feeling very comfortable with sending nudes, but she did it anyway. Various forms of coercion could have been present. If that were the case, the sexting would be unethical—not because it was sexually explicit material, but because it was nonconsensual or coerced.

Looking at sexting between adults, we might draw a few general conclusions about this form of self-made pornography. Sexting is often used as a way to affirm one's own sexual attractiveness and desirability. It is used as a sexual behavior similar to flirting or foreplay in a relational context. In order to have positive effects on the relationship, it must be consensual and wanted. While this is the predominant experience among adults, it is not universal. Some adults share a sext beyond the person for whom it was intended without the permission of the sender. However, the ramifications of this are often less detrimental than at younger ages. In particular, there can be special risks for individuals who have mental health issues or for whom sexting exacerbates gender inequality or violence in their relationship.[33] More on this discussion can be found in chapter 3.

Then what makes sexting so problematic for young people? First, let's dismiss the belief that sexual behaviors are detrimental to young people. In fact, consensual, planned, and protected sexual behaviors are not problematic and may help youth develop their sexual self-concept as well as patterns for future healthy relationships. Yet we know that even adults struggle with consensual, planned, and protected sexual behaviors. So those modifiers are important. Still, when it comes to sexting most young people do practice consensual, planned, and protected sexting to the extent that they can. That doesn't make it risk free.[34]

Let's deal with planned and protected first. Some young people report having a few sexually explicit photos of themselves already prepared to send if needed. The staging and curation of the photos is important, just like any other image a teen might post to social media. Teens also choose to protect their content to the extent possible both through technological means—using direct messaging applications for one-to-one communication or social media platforms like Snapchat that erase images after they are received—and through group solidarity. Some heterosexual girls show each other their sexually explicit images to receive suggestions on posing and to ensure that if images are forwarded, they are already semi-public among a sender's most supportive friends.[35] Of course this is necessary because of the risk that sexually explicit materials will be shared beyond the recipient—most likely, though not exclusively, among heterosexual males. In some cases, this has resulted in tragic social outcomes for young women.[36]

This brings us to the issue of whether teen sexting is consensual. There are few studies about sexting and LGBTQ+ youth; most research focuses on heterosexual youth, so a bias is present in adolescent sexuality research. In particular, gender inequality plays a significant role in sexting practices among heterosexual teens. One can imagine the consequences of outing a teen's sexual orientation or gender identity by forwarding a sext, but there is little research on this happening. Here I focus on the available research related to heterosexual teens. Sexting, like pornography use, is a practice that happens across racial demographic groups. The primary factors related to whether or not a teen sexts are access to digital technology and desire to be in or stay in a relationship.[37] The use of sexting is also different dependent on very particular circumstances. In heterosexual relationships, boys often request sexts rather than send them. A girl's choice to respond to the request reflects an assessment of whether she must send a sext to keep the relationship and whether it affords the boy sexual satisfaction.[38] In this case, the sexting is coercive. However, when girls send sexts or intentionally start conversations that lead to sexting, whether in relationships or not, their experiences are not coercive and often result in consensual sharing of sexts.[39]

Sexting, like pornography use, does not happen in a vacuum. Adults and teens contend with the sexual scripts promoted by the cultures in which they live. These scripts are often heterosexist and promote women sharing their images more often than men. But that does not also mean that women, even adolescent girls, are not creatively negotiating their own sexual agency, desire, and self-concept. In fact, we should note the ways that these sexual scripts harm male sexual development, as seen in the case of compulsive pornography use and sexual dysfunction. I am not equalizing experiences of exploitation against women and heterosexual male sexual dysfunction. What I am saying is the current state of heterosexual sexual scripts that focus on objectification of bodies and sexual arousal are harmful for everyone involved. That said, for many others sexting and publicly available pornography can improve relational ties and deepen their affirmation of sexual embodiment. Positive outcomes correlate with ensuring ethical features like consent, relational trust, and appreciation of bodies.

Embodied Theology and Sexually Explicit Online Materials

Danielle, Joel, and Peyton need a place to connect their use of pornography and sexting to their faith. Currently, that space is lacking. Most Christian congregations reinforce shame and universalizing moralism when it comes to sexually explicit materials. Instead, youth and adults fumble around on their own to figure out how to better understand the fullness of their embodied sexuality as part of God's good creation as well as live that out in relationships. In this chapter, we have discussed the need to share more widely the Christian theological foundations for the good news of embodiment—how we live out the image of God in each of us and cultivate attunement as part of our embodied ethical awareness. With this foundation, we can better evaluate the purpose and effects of using online sexually explicit material. Two important factors for consideration are whether the use of online sexually explicit material leads to or interrupts erotic attunement and how digital design features shape sexual engagement.

First, erotic attunement, related to the expansive understanding of sexuality, calls us to deepen our awareness of embodied feelings, desires, and emotions as well as honor each person we encounter as unique and equally shaped by their embodied feelings, desires, and emotions. As we engage sexuality explicit online material from a place of erotic attunement, we must see the humanity in the people appearing in an image or video. This suggests we should not consume videos or images in rapid fire succession or those that segment people into body parts or use close-up focus only on behaviors. Some will say it is impossible to connect with the humanity of or appreciate the feelings and emotions of those who appear in pornographic material. But the data seems to suggest that it is possible, especially for female-identified users. This also generates a relational awareness of one's connection to self and others, which can even be enhanced through digital means related to experiences of sexting and some shared pornography use.

Of course, this is difficult work, and not only because many of us carry shame and embarrassment fueled by sex-negative theologies. It is also because, while these technologies allow for ease of communication, access, and even affordability of content, they also perpetuate cultural stereotypes that can damage body image and sexual self-perception, providing digital avenues for harm. The ubiquity of pornography means users encounter "myths of sexuality, race, and male dominance . . . reproduced online."[40] Anonymity and availability mean that the internet houses a spectrum of material, from sexual behaviors that are degrading and violent to those that are alternative or out of the mainstream to basic vanilla sexual acts.[41] The mass amount of content and software designed to promote attention and response can generate compulsive use, though not among the majority of users. A behavioral effect of digital design that is far more likely is the sharing of self-made sexually explicit material or sexting beyond the intended receiver. The ability to save, reproduce, and spread digital content fuels both online pornography and sexting.

The reduction of sexual activity and sexual arousal to objectification and consumption purely through visual imagery has negative emotional, social, and physical sexual effects on the user. But using pornography for education and sexual empowerment deepens one's relationship to their body and enhances offline experiences of sexual arousal. The purpose for and duration of use—whether one is an adult or a youth—seem to determine positive or negative effects.[42] That is to say, in terms of attunement, there appears to be a way to use pornography to deepen one's understanding and attention to erotic feelings, desires, and physical sensations, as shown by the experiences of Danielle and Peyton. Joel, on the other hand, is stuck in a physical and emotional recognition of his erotic desires; but he resists a cognitive and social affirmation of them, potentially in response to his perceived ideas of bisexuality and heterosexism.

Jaco Hamman, a professor of psychology and religion, suspects online pornography use becomes more problematic for men for specific reasons. First, men, especially heterosexual men, tend to consume more online pornography. Because the brain is malleable and responsive, the "shortness of the clips demand clicking and searching for additional material to maintain interest and stimulation."[43] This leads to increased viewing. As noted in the discussion of embodiment, our bodies learn through practices and ritual. Hamman argues that viewing images has superseded sexual fantasy, which is slower, digs into memory, and connects to a wider embodied experience. The use of images through this rapid-fire digital visual means reduces the experience of arousal to a limited, external gallery. It habituates the user to approach bodies as objects when the encounter with them is spliced, transitory, fleeting, and entirely disconnected from relationship. Heterosexual men also tend to consume pornography that has fewer storylines and that often depicts bodies only partially.[44] These digital features diminish the fullness of sexuality and disconnect it from relationship.

Pornography content that is violent and degrading, along with nonconsensual sharing of sexually explicit material, is morally harmful. As theologian Miguel De La Torre states, "Sex may indeed be a personal (for example, masturbation) or interpersonal act. But it is not private. It does have public, social, and cultural ramifications."[45] Use of sexually explicit online material goes beyond moral judgments of personal use. We should advocate for the removal of certain content and create cultural expectations that it be produced consensually. We should not eradicate online pornography altogether because use of some of the material for certain purposes has not only personal, but interpersonal and social goods, including some forms of sexuality education.

Everyday Sexual and Digital Ethics for Pornography Use

- Cultivate body-positive Christian beliefs.

  Our bodies are to be valued and respected as part of God's good creation, and we can extend that to those we encounter via image or video online. Before sharing in intimate connections either offline or online, we should ask: Am I honoring my body or another's or using it as an object when I view pornography or participate in sexting?

- Consider your ethical standards for pornography use and content.

  What kinds of behaviors are depicted? Is there harmful sexism, racial stereotypes, or denigrating body messages in the materials? Is this consensual—honoring the feelings, emotions, and desires of those participating? Is this deepening a relationship with myself, others, and God, or is it alienating me from others and God?

- Talk about the technological and theological issues related to online sexually explicit materials.

  How can we push back against certain technological features? Am I choosing to view another image or video, or has the software made the choice for me? If the material is important to sexual pleasure, consider paying for curated sites that are explicit about their production and employment standards. How does sharing or forwarding sexually explicit material affect the person who created the material?

- Commit to practicing erotic attunement.

    We should seek to deepen our connection to our feelings, desires, and emotions—including sexual desire, attraction, and pleasure. What negative sexuality teachings are leading to shame about your sexuality? What theological reflections in this chapter help you affirm the _imago dei_ in you as a whole, complex, embodied creation?

I Sought One Whom My Soul Loves

TO SEARCH FOR AND FIND A HOOKUP, DATE, OR MARRIAGE

A T FIRST GLANCE, ONLINE DATING APPS do not raise many ethical issues. Of course, some of the apps meant to promote hooking up without any ongoing commitment might be seen as against certain Christian interpretations of ethical sexual relationships. Here the moral objection is primarily about sexual behaviors. Questions about how digital technology affects the way people meet and find each other to form relationships are a secondary or unmentioned concern. Yet Christians should raise questions about online matchmaking and relationship ethics. The majority of matchmaking apps take a person's community, such as friends and family, out of the process of relationship formation. Since communities of support have always been an integral part of Christian understandings of how to sustain and nurture believers through the ebbs and flows of their lives and relationships, we might wonder if this change is impacting how or why people form sexual relationships. Additionally, platform designs dictate communication patterns and what information is used to form connections. These designs may or may not align with values that support healthy relationships.

Two things are for sure: people increasingly meet online, and those online platforms determine how users present themselves and judge other potential matches. During the time of the global COVID pandemic, use of online matchmaking apps increased for obvious reasons, though this has been a steady shift even prior to physical distancing lockdowns. In this chapter, we will discuss a variety of factors that lead to an increase in use of online apps to meet people as well as the overall acceptance of these forms of relationship building. In nearly three decades, we have gone from the 1995 *X-Files* episode where a young woman is murdered by her first online date, to the popular Netflix reality show *The Circle* that chronicles all the ups and downs of contestants' dating schemes.[1] Meeting a romantic or sexual partner through a matchmaking app is commonplace.

We have come a long way in acceptance of—or maybe acquiescence to—using digital technology for sexual relationship formation. From the mid-1940s to the mid-1990s, most heterosexual couples met through intermediaries like friends. That practice has been declining since 1995, and sometime around 2017 digital connection took over the majority of matchmaking.[2] This makes sense given the overall increase in digital communication as well as generational shifts in use of social media and smartphones to stay connected across all varieties of relationships including friendships, family, and dating partners. Ultimately, users are primarily logging on to find connection, whether also for physical sexual pleasure or intimate sensual sharing. The need for connection on physical, emotional, and spiritual levels is part of being human; the way to find that connection is now digital.

Unfortunately, most online matchmaking apps are divided into two stark categories—dating with the hopes of a long-term relationship or hookups. This perpetuates an already concerning problem of either reducing sexuality to sexual behaviors or always directing it toward a fairy-tale forever coupling. These

two opposites—options that are more myth than experience—
are perpetuating harmful patterns of relationship and limiting
understandings of sexual flourishing. We need a richer and more
complex ethic of sexual relationship that honors the fullness of
our sexuality and intentionally pushes against harms like social
standards of attractiveness, confines of orientation and gender,
and sexual harassment. These patterns are not new, and with
more awareness (and incentive) digital platforms can be designed
with the potential for more creative integration of matchmaking
and community.

The Future of Digital Matchmaking

In his recent short story collection, Alexander Weinstein nar-
rates the relationship-seeking life of Mandy, the protagonist of
the story entitled "Comfort Porn."[3] The story opens with Mandy
attending a barbecue with her friends. The scene is an idyllic vision out of a young adult film in which each friend is more attractive than the other. Everyone is excited to see Mandy arrive.
Then the reader realizes that Mandy is using a virtual reality (VR)
application to attend this event and she doesn't know any of the
people. The "comfort porn"—socially scripted VR scenes—evokes
the user's feelings of belonging through fun, welcoming parties
or intimacy through long walks and sunsets on a perfect date. We
are told comfort porn can include dance parties, birthday parties,
and mundane walks outside, as well as sexually explicit material,
though Mandy usually pays for the friendship scenes.

   We soon find out that Mandy craves these interactions because
her world, in the not-so-distant future, is mostly cloistered because
of climate change and people do their jobs online from their private
living spaces. The primary way to meet people is the Firestarter
app, which has become the Amazon of dating apps. Dating has
devolved into sexual hookups. This app uses technology commonly
employed in current dating platforms—photos, bios, opportuni-

ties to show interest (by trying to light a spark), and messaging. The sexual encounters are transactional, with no expectation of anything other than sexual pleasure. The app includes safety and consent ratings based on how accurately actual encounters match what users list as behaviors they will or won't perform.

Other digital services in the story try to re-create the feel of in-person parties or the belonging sold in comfort porn, but most are staged. Mandy feels like the performative quality is akin to paying for friends, so instead she ends up joining a "hangout" group that sets up in-person meetings to play volleyball; they record the social scene and share it with others—like comfort porn. During one outing she meets someone in person, has a "comfort porn" kind of day of exploring and eating out, and has sexual intercourse with him that night. To her disappointment, nothing more comes of the encounter.

Mandy's occupation is editing profiles for users of Aphrodite, a less popular matchmaking site for those seeking dates rather than just hookups. Mandy critiques these users. In her view, they embellish their descriptions and try to justify their need for sexual relationships under the guise of dating. Using Firestarter means, from her point of view, she doesn't need to pretend that she wants anything more than sex. But we already know from her love of comfort porn that she wants friends and belonging. When an advertisement pops up on her social feed for a free air conditioner, a coveted item when living in an apartment with no air and temperatures of 120 degrees, she needs to find a person who will help her move it. She comes across an old email from a college friend, Katie, and decides to reach out to her. Katie lives outside the city and, according to Mandy, is weird for randomly suggesting an in-person visit. Katie turns out to be a heavy-set, middle aged woman who got married, then divorced, survived brain cancer, and got off social media. Now she lives a modest life with a boyfriend and teaches art classes as an adjunct. When Mandy meets Katie, she at first negatively judges her appearance, life outside the city, and less-than-attractive boyfriend. But within a day, Mandy envies Katie, who lives the life that comfort porn projects—one with relationships filled with belonging, intimacy, and meaning.

In this short story, Weinstein captures a possible trajectory of matchmaking that is overwhelmingly confined to online platforms amid wider cultural shifts—changing social, political, climatological, and economic structures that (like COVID) may someday force us all to connect primarily through digital rather than physical means. He exaggerates current aspects of online dating and sexual encounters to magnify their effects on relationship formation and human flourishing. We see the overreliance on appearance through photos; on algorithms that create matches; and on platform protocols like bios and ratings, scripted forms of communication and interaction that find their way into in-person recorded "hangouts." We also sense the feeling that since everyone is doing it, there is no alternative. While Mandy (and presumably others consuming comfort porn) wants something more, the heterosexual men that Mandy interacts with appear to enjoy the social landscape that Firestarter creates. Gender differences in online matchmaking are common; users' sexual orientations and gender identities directly shape the way they engage platforms and dating more generally (more on that later). We might also imagine a future where social institutions like family, the state, or religious groups return to a system of matchmaking that is devoid of personal choice. This would also reduce partnering to sexual behaviors for the sake of procreation rather than pleasure, maximizing the state or community need for certain offspring. Neither that world nor the world Weinstein conjures is fully upon us, though we can see fragments of both active in current social practices of matchmaking. As Katie's presence demonstrates, no social shift caused by technological changes becomes absolute. However, it often becomes dominant or normative in a way that shapes our imaginations and future possibilities. Research suggests that online dating is slowly creating a more monolithic global dating culture.[4]

Online Matchmaking

What is the current landscape of matchmaking applications? Most adults coming of age use online dating apps. Among all US adults, close to one-third have used a dating site. Use is affected by things like relationship status, age, geographic location, and sexual orientation.[5] For example, close to 50 percent of people who are eighteen to twenty-nine in age or who identify as lesbian, gay, or bisexual have used a dating app. While some think online dating apps are only for hookups, Pew Research found that one in ten users end up with a long-term relationship or marriage, and it's higher for gay-identified users. Most users feel safe using dating apps, though there is a high level of sexual harassment experienced by some 60 percent of young women on dating apps. This is perhaps why Weinstein adds the safety and consent user ratings in his futuristic dating app.

While use of online dating apps is on the rise, they are not all the same. Each application has different features, platform requirements, and even specifications of users. An example is something like Bumble, which is geared toward heterosexual female user empowerment, where only a female can initiate a conversation and racist and heterosexual harassment results in guilty parties getting kicked off.[6] Other applications limit participation to users based on race, age, religion, or interests. Examples of these include BlackPeopleMeet.com, SilverSingles, FarmersOnly .com, ChristianMingle, and JDate.com. Also, "in the case of LGBT communities, online matchmaking sites offer new possibilities of communication without the danger of stigmatization that interactions offline can entail."[7] Sites like HER or bearwww.com are closed communities to provide safety for LGBT users, and many apps build in multiple ways to identify sexual orientation and gender identity that are usually static categories on other sites.

When it comes to a user's purpose, some larger dating sites specifically market to users looking for long-term relationships, but it doesn't prevent users from going on only one date or meeting to hook up. Sites promising a long-term match usually require significant amounts of personal information to promote their algorithmic matching capacities, including personal interests, demographic information, and preferences related to relationship qualities. That said, many wrongly assume that Tinder or Grindr are only hookup sites; users often negotiate a range of intended meetings, from one-time to multiple-time hookups, dating, or long-term coupling. What dating sites and apps have in common is that they work as mediators, just like friends and families did in the past.[8]

The different user groups, purposes, and features of matchmaking applications or sites combine to determine communication patterns and relationship formation. Internet researchers find that "Profile scripts, social network norms, and regulations guide users' self-presentation and provide a framework of pre-established relationship patterns within which to perform their search for the ideal partner."[9] Even with site-specific predetermined methods of communication, researchers consider these to be spaces of identity formation and negotiation. The platforms often allow for curation of information in ways unavailable during in-person introductions. In the short story, Mandy's occupation is to help users of a particular site craft their profiles and consider, for example, how certain photo content elicits more responses. She uses similar skills to select genuine users instead of spam accounts on Firestarter or swipe past misogynistic profiles with "dick pics" as their main photos. Identity curation happens within the parameters of the platform. Some sites require more photos while others rely on bios or screening surveys. Some would argue knowing certain forms of information about someone before making a match with them increases compatibility, but it may also allow external factors to have an outsized influence.

Both online and offline matchmaking practices have shared emphases such as, to name a few, physical appearance, shared interests, demographic similarities like education level, geographic location, and income or professional levels. Offline cultural standards very much influence curation choices. For example, dominant standards of beauty (light-skinned, thin, athletic, and young) shape users' expectations and choices in matchmaking. Even when physical appearance is removed or minimized as a feature, other social markers like occupation, education, and hobbies come to signify desirability. The science of matchmaking has always been influenced by social, political, and economic forces. Sociologists call this phenomenon homogamy—meaning, globally, people tend to seek a partner who shares similar social and cultural characteristics like religion, race or tribe, education, age, and economic status. In the past, families and geography were major factors in maintaining homogamy. However, that is no longer true when a person has a pool of hundreds if not thousands of possible online users with whom to connect. And yet, homogamy generally persists, especially in long-term partnerships. For example, eHarmony reports that its matchmaking algorithm relies on the study of what predicts marital success, primarily "core personality traits and key values."[10] Personality traits may be unique to individuals, but key values are overwhelmingly shaped by the social factors named above like religion, race or tribe, geography, education, and economic status. Online matchmaking projects the possibility of a diverse landscape of potential matches, but in fact platform design relies heavily on appearance and algorithms that reinforce social status likeness.

Bias based on social factors is not only a design issue. Many users simply may not have considered their own preference biases. Some describe these biases as subtle and unconscious.[11] So users bring their own biases with them, and often platforms are designed to exploit or reinforce those biases. For example, because users make assumptions about disability and ableness on sites, some people with disabilities preferred explicitly labeled "disabled-only" sites so they don't have to worry as much about discrimination or disclosure. But that also can lead to the assumption that all other sites are "disability free." In light of all this, many people with disabilities are extra careful about how they describe themselves to avoid the discrimination they often face.[12]

Overwhelmingly, these applications contribute to judging compatibility based on appearance or algorithmic schemes that reinforce traditional social markers like race, economics, education, and so on. The promise of online matchmaking that overcomes traditional offline social boundaries is false. Of course, there is something to be said for building a match based on shared values and experiences, especially if one is from a minority community. Some dating apps cater to a specific demographic like those mentioned above based on religion, occupation, or sexual orientation and gender identity. The closed community affords a certain level of safety in some cases and a predetermined factor of compatibility. However, each of these platforms reduces the ever-present diversity that is part of all demographic communities. For example, many Christian matchmaking apps promote only one theological and demographic slice of Christianity— predominantly evangelical, conservative, and white.

A plethora of options for potential online matchmaking has also raised questions about what Dan Slater calls "choice overload."[13] Slater suggests that the sheer volume of options results in unsatisfied users, leading them to keep searching for something else, simply because they can. Yet the data doesn't point in this direction. Rather, those who meet online and make a match, especially those wanting a long-term partnership, tend to stay together. What about those who never make a match—those who spend hours swiping, liking, and sending initial introductory messages? Kate Julian discusses their situation in detail in her article "The Sex Recession."[14] She chronicles the current adult response to online dating and hookups, noting that the gamification of dating apps with a myriad of options leads users (especially heterosexual women) to be choosier and to develop a feeling of futility when hours are spent scrolling and little or no connection is made with a match.[15]

Julian notes that the shift from meeting partners in person to using online apps is influenced by and reciprocally affects factors related to sexual practices and relationships.[16] Contrary to popular hookup culture myths, not everyone is engaging in sexual behaviors all the time with random people they meet online. Her reporting suggests a few factors are influencing the decline in sexual behaviors and the use of dating apps. The first is the rise in masturbation practices, or sex for one. Young people don't need a partner for sexual pleasure. The second is the reality that teens and young adults are overscheduled and future oriented. Sexual intercourse and long-term relationships do not fit into their life plans, so they delay them. Third, she suggests bad sex is the result of poor sexuality education and the ubiquity of online pornography, especially affecting males in heterosexual relationships.[17] Fourth, social practices lead to greater inhibition around nudity, and this is mixed with a culture esteeming perfect bodies and social media tools (like Snapchat filters) that erase physical imperfections online. Inhibition can also be the result of body shaming or negativity exacerbated by remnants of purity culture messages taught to Christian teens and tweens, especially for women.[18] Last, Julian delves into what she calls the "Tinder Mirage."[19] She wonders why people don't just give up after failing to find a connection using a dating app or scrolling through hundreds of profiles a week without sending a message. What she finds is that everyday, in-person communication is seen as strange. Some young heterosexual men report that in a post-#MeToo climate they are unsure how to proactively initiate a date in person. The difficulty is only exacerbated when they turn to pornography for sexuality education and relationship models, as discussed in chapter 1. Julian's research is helpful to provide a view that looks beyond simply how platforms or digital design affects online matchmaking. We need to better equip people with sexuality education and a body-positive sexual self-concept.

Important to the issue of online dating, we should be equally concerned with the ways digital communication not only requires critical digital literacy—like Mandy's ability to discern profile information and what it says about a user—but also replaces in-person communication skills. It's true that we are all far less inclined to talk to random strangers and would rather stare at our phone to occupy free time in public. Yet the bedrock of healthy relationships is comfort with one's body, healthy communication, and communities of accountability. How do we find that in a world where matchmaking communication is one-on-one digitally mediated images and text? There may be ways online communication could increase offline honesty and dialogue. For example, it might be easier and safer when negotiating the use of certain platforms or accepting certain requests to ask a potential match about sexual history or to establish the purpose of meeting offline (is it a hookup, a date, a step on the way to marriage, an opportunity for companionship?). This is because the technology itself encourages certain forms of communication, as depicted in the short story discussed above. The Firestarter user knows exactly what a fire emoji means, for example, and they understand that by selecting a match and starting a spark they are agreeing to meet in person.

What's the balance? It seems most people, not just those on online dating apps, have similar questions about the impact of online formats versus in-person communication. Add to that the mythology of a hookup culture versus an old social (and prominent Christian) script that all dating is for the purpose of securing a marriage partner, and I fully understand the desire to opt out entirely. We need to better understand how different forms of communication have different advantages for relationship building and stop engaging in those that do not. And we need a Christian approach to relationship building that promotes shared worth and value, and that recognizes—even supports—the fact that relationships require work and change over time.

Relationship Ethics

We are stuck between two relationship myths which are socially defined by a Christian stereotype of marriage that has never existed. The original myth is based on a recent, not ancient, ethic of no sexual intercourse until marriage. The Western Christian assumption that couples will wait until marriage to engage in sexual intercourse is only a few hundred years old.[20] Globally, this is not a monolithic Christian practice or teaching. Yet most US Christians labor under the illusion that one is sinful if they have sexual intercourse prior to marriage and that marriage is the only type of relationship in which one can engage in sexual behaviors. This leads to a false ideal that all sexual and romantic relationships should be judged against or moving toward. This forever-marital match fits a particular economic, gendered, and reproductive picture. Rather than unpack this historically recent, exclusionary ideal, many in the West have fallen for another myth, the hookup. If sexual behaviors are no longer confined to the goal of marriage, we can divorce sexual behaviors from any ongoing relationship. Most people aren't having sex with random strangers they meet online, just as most people aren't waiting for marriage to have sexual intercourse. The key data point is *most people*. The two dominant relationship ethics—abstinence only until heterosexual marriage or no relationship needed for sexual behaviors—do not reflect the reality of most Christians' lives.

These myths are also harmful to the flourishing of sexuality in a just and inclusive way. They do not honor the fullness of humanity's sexualities created by God. The first myth suggests that in order to engage in sexual behaviors one must be heterosexual and, in some traditions, always open to procreating. Then wrapped into that Christian ethic is a dose of gender dynamics that relegate men and women to particular roles regardless of their charisms or vocational desires. If you are a woman, you serve your husband first, then hopefully children second; these are to be your dreams and desires or you are going against God. In the United States, the creation of this heterosexual marriage myth is an unholy alliance, a melding of certain narrow Christian interpretations deployed to support white economic structures.[21] Some Christian ethicists have argued that the movement for marriage equality and gay rights was shortsighted in its desire for inclusion into a flawed system. The need to secure marriage rights for medical, employment, parental, and legal benefits is a perfect example of how Christian understandings of marriage are more about maintaining a certain social order than supporting sexually holistic and flourishing relationships.

One might argue that hookup culture dismantles the exclusionary gender and orientation practices found in the aforementioned stereotypical Christian understandings of marriage. That may be true. Removing sexual behaviors from a predetermined relationship standard promises egalitarianism across genders and orientations, focuses on sexual pleasure, and frees people from compromises to their life goals regarding education, geographic relocation, and job seeking. Yet the reality of hookups, and hookup culture more generally, does not bear this out.[22] Findings suggest that sexual pleasure is usually not shared or equal, and in heterosexual encounters women rarely feel sexually satisfied. People in these encounters have higher levels of substance use. Maybe this masks sexual discomfort and reduces communication; definitely this affects consent. The myth of freedom from commitment does not equate with the reality of how most participants feel following hookups. Survey respondents share feelings like regret, shame, or confusion about wanting a relationship but not knowing how to move beyond a hookup. These experiences are not unique to heterosexual encounters either. Research suggests that LGBTQ young adults more actively critique hookup culture; yet they also struggle with experiences of communication and consent or exploring a wider range of pleasurable behaviors in their own hookups.[23]

Whether it is no sex until heterosexual marriage (celibacy for the rest) or sexual behavior with no strings attached, these sexual ethics do not generally produce healthy, inclusive, holistic sexual relationships and should not be the primary shapers of online matchmaking applications. Both myths are contrived in ways that serve economic and gendered agendas that are oblivious to who we really are, to what sexuality really is, and most importantly to the kind of world that promotes love and justice across all relationships. Marvin Ellison, a Christian sexual ethicist, says it best when ruminating about why queer Christians shouldn't settle for marriage equality:

> Celebrating our common humanity requires making an odd, decisively queer turn toward radical equality and plunging in together to rebuild a vibrant, just, and wildly inclusive social order. . . . The queer agenda has never been only about sex or even sexual justice, but rather remains a persistent, unwavering demand for a comprehensive renewal of life-in-community. The change we desire, deep down, is not mere inclusion but rather spiritual, moral, political, economic, and cultural transformation from the grassroots upward and from our bedrooms to far beyond.[24]

That is to say, whether one identifies as queer or not, very few of us fit into or are even happy with the current options available. The system of marriage in the United States that was created based on and seeking to reinforce gendered, economic, and racial divisions is no longer tenable for most Christians, if it ever was. Hookup culture is equally disappointing and problematic. Significant social, economic, and technological shifts make the use of online matchmaking apps a useful and even necessary part of dating practices. How do we remake or deploy them in creative ways that allow digital connections to contribute to "spiritual, moral, political, economic, and cultural transformation" in our bedrooms and beyond?

First, we need to know what we are looking for, then we need to provide sexuality and digital education to get us there. If Weinstein's protagonist, Mandy, is any indication, three things come to mind as central concerns among those who turn to online matchmaking: companionship, clear and honest sexual communication, and mutual pleasure. Sadly, for Mandy it seems companionship has been separated from sexual pleasure, and clear, honest sexual communication is reduced to a checklist of sexual behaviors. For some users, the detachment of sexual pleasure from ongoing companionship and clarity about behaviors, including a feedback loop to report safety and consent issues, remedy some negative aspects of traditional Christian sexual standards. However, they do not go far enough and certainly do not promote holistic sexuality, which requires attentiveness to intimacy and sensuality as aspects of healthy sexuality.

The desire to be loved (beloved)—connected, cherished, desired—has been skewed by a heterosexual, procreative culture, a theology built on gender inequity and fear of sexual desire, and the subsequent response, which has divorced spirituality from sexuality in hookups. Taking a cue from the Song of Songs, we can affirm that Scripture recognizes the beauty and value in erotic, embodied sharing between two people. Some interpreters want to erase the sexuality and desire present in the text by reading it as an allegory between God and Israel or suggesting the two people in the text are married, but many scholars dispute these readings as there is no solid textual evidence for either.[25]

Against all odds, this text made the cut when Scriptures were chosen to comprise the Bible. This testifies to the importance of its message.

The protagonist identifies herself as black and beautiful (1:5) from laboring in the sun (1:6), a mark of her class status and independence (1:7).[26] These identifiers could be read with the same racist and classist judgments that oversexualize women of darker skin and lower class today. Instead, the text sets her up as an example of empowerment against these factors. She is not alone in her desire, as a group, the daughters of Jerusalem, join in chorus with similar desire offering the refrain of the text. While bridal and wedding imagery and language are used, the two do not appear to be married, and there is no conclusive data to prove that they are.[27] In addition to desire, the text acknowledges the risk the woman takes in seeking her lover because of the dangers of other men and resisting her defined role. As a reader moves through the chapters, the poem (or maybe poems) can be disjointed and lose a coherent timeline and narrative. But that's precisely the beauty in raising them up for scriptural purposes. This is not a dos and don'ts list of sexual practices or ethics.

One reading of the Song of Songs is as testimony to how social confines cannot limit the flourishing of erotic desire embodied in human relationships. The Song of Songs "presents an unusual view of premarital sexuality and non-reproductively focused female sexual initiative. Moreover, the jealous, overpowering male of many other biblical texts is replaced here with a male lover willing to be vulnerable to love, to yield to it."[28] While the text reinforces heterosexual attraction, it challenges other dominant systems like family, gender, economics, and race, to name a few. This is in line with the challenges that Ellison supports and that I'm suggesting are necessary as we rethink an ethic of relationship. The Song also promotes a spiritual connection with erotic desire toward creation and the divine, locating the sacredness of sexuality in both. Even as the lovers stand against social norms, there is an ever-present community in the "daughters of Jerusalem" that keep the refrain.

In the introduction to this book, I share two key aspects related to sexual ethics. The first is erotic attunement and the second is a values-based approach to sexual relationships. From the Song of Songs, we can see how Scripture acknowledges the role of erotic desire in forming relationships that allow for the flourishing of each person rather than reducing them to the object of the other. The mutuality in this depiction is not one-to-one equality. Mutuality means seeing the other person for who they are and desiring the fullness of what they can be. Audre Lorde describes eros, passion, and power as that "which comes from sharing deeply any pursuit with another person."[29] This could be sexual or not for Lorde. In fact, Lorde stresses a creative generativity that comes from the sharing. An erotic exchange is meant to be creative. Companionship—like what Mandy longs for or what I would reframe as accompaniment and friendship—is not absolutely necessary to yield mutuality and creativity in a relationship, though it may make it easier and sustain it longer. This requires time and communication.

Many of us experience a level of awkwardness rather than awe when communicating about sexual attraction and desire. We are not poetic, or even confident, like the lovers in the Song of Songs. This is partly because sexuality has been treated as a taboo subject and not connected to the rich, sensual imagery and language of religious ritual and faith practices. We need to connect sexuality and spirituality more readily so we can develop a vocabulary that speaks of sexuality and sacredness without hesitation. Additionally, when we consider the labor of relationships, we need to readily see the confining forces—like economics, race, gender, orientation, and so on—that limit mutuality. Like the lovers in the Song, we can push beyond these limits using the erotic as a creative power. Relationship then becomes a connection not only between people but also with the divine and all of creation. David Carr in the conclusion to his chapter on the sexual and spiritual connection in Song of Songs writes:

> On one level are our attachments to loved ones, nature, and other things. It is easy to be caught up in that level alone, and become addictive and destructive. Yet there is also the possibility of seeing our world through another transparency, where our beloved, our world, shimmers with the reality of God. On one level, eros opens us to sensuous connection with another person, a poem, a piece of nature, the world. Yet on another level, our eros opens us to experiencing God loving us in and through those things.[30]

When we connect not only to another human but also with God in and through that erotic relationship, the values of mutuality, creativity, and companionship or accompaniment are realized. We do not need marriage for this; we need honest communication and sustained presence. A presence that likely needs more time to flourish than a hookup allows, though it might start with a hookup so long as both are open to this potential (which as we see from research, might not always be possible, depending on participants' intentions).

Relationship Ethics for Online Matchmaking

Can online matchmaking applications promote mutuality, creativity, and accompaniment? They might be able to, but first and foremost users need to come with the intentionality and awareness that sexuality and spirituality are intertwined. Thus, whether online or offline we should not "use" another person—participating in exchanges that ignore the fullness of one's self or that of another person. This can be difficult when a user just wants to find a match. We have to admit there may be a good deal of trial and error in the process. Those challenges suggest we need a community of accountability that is found beyond the matchmaking online application population. We need friends and community with whom to discuss our choices and experiences.[31] Online matchmaking can cut off the influence of friends and others on our decision-making. This can also lead to an assumption that other social forces aren't influencing matchmaking, and the data shows this is not true. User intentions and communities of support are a key part of healthy sexual relationships.

Discernment about which apps one uses and how one uses them is another key factor in cultivating mutuality, creativity, and accompaniment. If the primary use of an application is for one-time hookups, users approach matches without a commitment to sustained presence and likely won't share helpful information that leads to mutuality. On the other hand, if an application says it can find you a marriage partner based on an algorithm, users should be informed about how the algorithm works and about the fact that sexual desire does not automatically follow a mathematical equation of common social statuses. There are a variety of matchmaking applications that promote users getting to know one another on a variety of different levels—including the use of pictures indicating physical attraction, longer bio descriptions, chat and messaging services that promote discussion rather than quick hookups, and so on. Applications can promote ways of interacting that are similar to how people build friendships. Those that combine these with the explicit acknowledgment of sexual attraction are more likely to yield matches for the purpose of companionship, mutuality, and creativity. Some designers, like those behind Bumble and HER (mentioned above), are even working to reduce gender, orientation, and sexual harassment issues.

Sexual pleasure, erotic desire, and partnering create a mix that is seductive because they take effort, and then come with ease while still requiring active nurture. These experiences constantly call forth in us new ways to be and to witness God in our midst. If we measure every person and profile as a potential marriage partner, we will miss the erotic that is present across a variety of relationships. If we have to be drunk or high to follow through on a match, we have been desensitized to the erotic in ourselves and the other. If we don't see the divine in the person in the profile picture, using them for self-pleasure (coerced or consensual), we have dehumanized the other. We are far too valuable and capable of erotic love and healing to live by sexual ethics that dehumanize us. Sexuality is sacred. "Seeking one the soul must have" is a testimony to the desire for erotic union based in sensual and intimate affirmation of one another. In order to achieve that, sexuality must be expressed and invited with mutuality, creativity, and companionship.

Everyday Sexual and Digital Ethics for Online Matchmaking

- Write your own sexual ethic before you start or continue with online matchmaking apps.

  The approach of "I'll know it when I see it" isn't helpful. What are five core values that are most important to you?

- Consider helpful theological or scriptural references for sexual flourishing.

  This chapter turns to the Song of Songs as a positive example of eros and sexuality. What other texts can you find? What ritual or prayer practices enhance intimacy or connect with sensuality? Texts or practices that denigrate the body or discriminate based on gender or orientation are harmful and should be avoided.

- Compare matchmaking applications before selecting the ones you will use.

  What is their main purpose? What do they allow you to share about yourself and learn about other users? What types of communication do they allow inside the platform?

- Talk to friends who use online matchmaking applications about your shared experiences.

  Even if friends don't serve as the conduit for blind dates anymore, it doesn't mean they shouldn't be part of the process of decision-making. Friends are a resource for support. How are their experiences alike or different from yours? What are their reflections about this process?

- Question the biases you bring to matchmaking.

  What standards of beauty are your default? Is that excluding you from meeting certain people? What status markers (like profession, education, or geographic location that drive algorithms) do you think should be avoided?

- Look for positive models of communication in a relationship.

  This might be a friendship you have or a family member's relationship. Get a little scientific and dissect what makes the communication work.

Love Does Not Delight in Evil

TO SURVEIL, EXPLOIT, OR VIOLATE ANOTHER

DIGITAL TECHNOLOGIES CAN BRING PEOPLE TOGETHER. They can also be wielded to surveil, exploit, and violate others. For example, the use of "cheating" applications, platforms, and devices increase the risk of intimate partner violence and abuse. When a user logs into LoveIsRespect.org, a healthy relationship and violence prevention site for young adults, a pop-up window notifies the viewer:

> Heads up! Your browser history can be monitored without your knowledge and it can never be wiped completely. Think your internet use might be monitored? Call us at 1.866.331.9474. Learn more about staying safe online and remember to clear your history after visiting this website. You can quickly leave this website at any time by clicking the "X" in the top right or by pressing the Escape key twice.[1]

These types of safeguards remind users to be cautious about their online data trails. A regular feature of digital technology, like search history, can be used by an abuser to track another person. In most cases, the software and hardware used to do harm were not designed or intended for this use, and yet human sinfulness repurposes them for abuse, harassment, and control.

Technology specialists often talk about affordances and constraints of different technologies. This simply means that each technology has certain benefits or capacities as well as limitations. For example, one of the main affordances of social networks is the capacity to make information go viral. The software is built so that one post can be seen by thousands if not millions of people by leveraging aspects like use of hashtags, size of one's network, content popularity algorithms, and paid boosts to content. These affordances are what make social media platforms great spaces for advocacy and engagement. However, the same capacities or benefits can be used for negative goals like sharing an ex-partner's nude pictures or slandering their reputation in another way. The extent of the harm increases based on the reach of the technology. The technology does not create sexual violence or abuse; rather, the impulse to harm, harass, or surveil comes from a human desire for power and control.

Condemning harassment and violence appears at first glance to be an easy area for ethical agreement, but what constitutes violence is not always easily agreed upon. Is one person's desire out of "love" to know everything about another person stalking, manipulation, and control, or is it information gathering and compatibility research? If someone asks a partner to remove a social media picture in which she is sexily dressed and striking a flirtatious pose, is that controlling or protecting her image? Is using a GPS locator app to follow a loved one for safety or surveillance? Is that concern or distrust? Each of these examples can have good intentions behind them, but they do not necessarily live out Christian values grounded in the love commandment, such as honesty, mutuality, or respect. When a user scales up and does these things repeatedly, they can inflict trauma, and there can be legal implications.

Living out values in a relationship requires a significant amount of communication as well as erotic attunement. Knowing one's own feelings and desires empowers each individual to love themselves fully, or at least strive for that. Too often, other things—like jobs, social privileges, or beauty—and other people— like spouses, children, or friends—determine our worth. Christian teachings of self-sacrifice and enduring violence for the sake of a greater good can confuse believers into thinking they should be suffering. They think, or are directly told, that the emotional, spiritual, or physical abuse they are suffering is their fault because they aren't sacrificing enough. These theologies perpetuate further harm against victim-survivors. Christian moral imagination often falls short of conceiving of a world free of violence. It is long past time that faith communities promote spiritualities of resistance to sexual violence, harassment, and abuse online and offline.

Christian Theology and Sexual Violence

I admit that after years of writing and teaching about sexuality education, sexual health, and violence prevention in faith communities, I am frustrated by the continual need to address the way Christians misuse theological teachings to perpetuate intimate violence. It's wrong, it's wrong, it's wrong! And yet, child sexual abuse, dating violence, and domestic abuse are prevalent in Christian contexts. Sometimes I think Christian theology and faith communities do more harm than good. I wonder whether advocacy hours are better spent on educational and economic empowerment for those who disproportionately suffer abuse and harassment, including women, children, and queer and trans people. And then I am reminded by colleagues and friends who are victim-survivors that their Christian faith sustained them in times when they almost didn't survive and often gave them the strength to break free from violent and abusive circumstances and relationships. Christian teachings and communities can be sources of justice and nonviolence if we are willing to root out and stop repeating centuries of abusive theology.

Many Christians have been taught to interpret Jesus's dying on the cross as a self-sacrifice and chosen suffering that Christians should admirably model in their daily lives. In other words, being "Christlike" means suffering and sacrificing for those around us. If people put their own needs and desires first, they are seen as selfish and unchristian. Somewhere in the historical trajectory of this theology, starting with the Pauline letters, these notions of self-sacrifice and suffering were given a gendered twist. In a heterosexual context, men sacrifice for the sake of serving God and women sacrifice for the sake of serving their husbands. Atonement theologies—the teachings that seek to make sense of Jesus's death on the cross and resurrection—have also been used to justify things like suffering and sacrifice associated with colonialism and enslavement to present day racism. Throughout history, when a particular Christian social group wants to justify its power over other people or peoples, they proffer that just as Jesus suffered, so suffering is required of good Christians, whose sacrifices will be rewarded in the afterlife. Of course, the suffering and sacrifices are for those being oppressed rather than for the oppressors, who claim they are chosen leaders and have to help the *other*, lesser human beings make sense of and be glorified through their suffering. For centuries, liberation theologians have likened people's suffering under economic, gender, and racial oppression to ongoing crucifixions.[2] Traci West, a Christian ethicist and advocate for ending sexual violence, writes that "this merger can create forceful cultural sanctions that enable intimate abuse and violence to proliferate, at great cost to victim-survivors, especially socially marginal ones."[3]

What do these teachings look like in everyday faith community experiences? Victims, especially female victims of male domestic violence in Christian marriages, report being told by pastors and priests that they needed to stick it out, change their behavior so as not to anger their partners, and offer up their suffering to God for the sake of the marriage. This represents one of the most stark and egregious misuses of atonement theology related to intimate violence. More banal uses equally deform the theological weightiness of Jesus's death and resurrection. Some youth group leaders might use Jesus's sacrifice on the cross to bring a bit of perspective to the drama of teen dating. For example: "Jesus gave his life for us; how can you spend so much time worrying about who will take you to the dance?" The message suggests that concerns about dating relationships are petty in comparison to Jesus's sacrifice. But then, how does a teen know what rises to the level of theological concern in a dating relationship?

In the United States, April is domestic violence awareness month. It is also the month in which Christians most often celebrate Easter, a joyous celebration of Jesus's resurrection after being publicly imprisoned, beaten, and crucified.[4] The version of atonement theory mentioned above suggests that Jesus's gruesome and violent crucifixion substituted for human sin, thus freeing humanity from God's wrath. The message assumes violence and abuse is a necessary requirement for the positive outcome. From a Sunday school teacher helping teens consider how faith informs their relationships, this message can cause confusion and even harm. Those who have experienced and are experiencing intimate violence may feel, or have been directly told, that they should suffer their violence in silence or they should lift up their suffering to God *just like Jesus did* on the cross.

The valorization of Jesus's death can justify suffering of others in violent contexts. Marie Fortune, in *Sexual Violence: The Sin Revisited*, has pointed out that Jesus's suffering is *voluntary*; he chooses to be crucified. Contrary to that, individuals suffering domestic and sexual violence are *involuntarily* suffering; they do not choose to be abused.[5] In direct opposition to the penal substitutionary atonement theory—Jesus's death substitutes for the punishment due for human sins—womanist theologian Delores Williams makes the argument that Jesus did not need to die and his death on the cross does not save us.[6] Instead, attention to Jesus's life, ministry, and teachings is what is needed for salvation. If intimate violence is involuntary suffering and we are to learn from Jesus's actions and teachings, how should we respond? In today's violence prevention language, we should be upstanders (stand up and do something) rather than bystanders when we see signs of intimate violence in relationships. Jesus's own teachings about the love commandment remind us that no one has the right to treat another person in sexually, emotionally, or physically abusive ways. While violence is part of the Christian story in a variety of ways, when we celebrate this violence and certain theologies that surround it, we silence those who experience various forms of violence.

Another crucial bit of theology contributes to justifications of sexual abuse and harassment—or at least to dismissals of punishment and quick moves toward forgiveness without accountability. Feminist theologian Rosemary Radford Ruether shifts the origin of sin from a biologically transmitted brokenness to an "inherited, collective, historical dimension of sin."[7] Yes, we are a sinful people, but not because of some genetic predisposition inherited from Adam and Eve, as some theologies would like us to believe. Rather, sin stems from social circumstances and choices that move us further from relationship with God. Through this understanding, Ruether redefines the theological concept of metanoia as a "process of emancipation" from oppressive social and cultural systems "to create a new self and a new society."[8] I have written about her concept of metanoia related to other aspects of digital technology, especially the way digital surveillance and memory function to keep records on everything users do online.[9] In her view of sin, Christians have individual human freedom and accountability, but she recognizes that moral choices do not exist in a personal vacuum.[10] That is to say, a wider culture of Christian theology that valorizes self-sacrifice and suffering contributes to social circumstances that make it difficult for victims to resist and escape the abuse they suffer. When these theologies are connected to racial, gender, and economic discrimination, it is that much more difficult for marginalized people to escape ongoing violence and sexual exploitation. Metanoia requires community response and accountability, not only individual freedom and forgiveness.

As part of metanoia, Christian faith communities need to realistically and courageously assess their roles in contributing to gender and sexual violence through theologies that perpetuate sexism—men's dominance over women—and an essentialist gender binary—the claim that two biological sexes perfectly align with gender norms and roles of maleness and femaleness. One does not have to look very far to find examples of conservative Christian teachings that claim homosexuality is an abomination and that God created women to be subservient to men. Many Christians misuse the creation narratives in Genesis and other Scripture passages out of historical context to support these forms of discrimination. However, the way Christians use Scripture to defend prejudices says more about their Christian values than it does about sound scriptural interpretation.[11] Christian ethicist Marvin Ellison suggests that "a patriarchal ethic grants permission only for those erotic exchanges in private that uphold the social hierarchies of male gender supremacy and white racial dominance." And "the traditional Christian sexual ethic is implicated in this mess because it has legitimated an ethic of male entitlement over women and female bodies" and over those men deemed submissive or inferior.[12] Sexual ethics becomes about power and domination, not about erotic desire or healthy sexuality.

One modern, digital example of this form of male dominance and hatred toward women can be found in the "incel" movement. Incel is short for "involuntarily celibate." In her article, "The Rage of the Incels: Incels Aren't Really Looking for Sex. They're Looking for Absolute Male Supremacy," Jia Tolentino chronicles the rise, beliefs, and impact of the incel movement. She writes, "In the past few years, a subset of straight men calling themselves 'incels' have constructed a violent political ideology around the injustice of young, beautiful women refusing to have sex with them. These men often subscribe to notions of white supremacy. They are, by their own judgment, mostly unattractive and socially inept. (They frequently call themselves 'subhuman.') They're also diabolically misogynistic."[13] These mostly young and white men do not fit the attractive heterosexual cultural stereotype, but instead of pushing back on the toxic masculinity that shapes this stereotype, they decide to make women—particularly young, attractive women—the targets of their rage. They believe they should be able to have sex with them whenever they want, and they blame movements like feminism for promoting women's right to control their sexuality and even for promoting body positivity for all people. They most often stalk women online and harass them through direct messaging on meetup apps in addition to slandering them on public social media accounts. This is a serious matter. A number of followers of this movement have done physical harm offline, carrying out multiple shootings and suicides.[14]

Tolentino reminds the reader it is men who have created the incel predicament and male ideals—as opposed to female or gender-inclusive understandings—that perpetuate patriarchal and heterosexist fantasies. Some might argue that there is a significant chasm between teachings about women being silent in church and serving their husbands, and incel beliefs in sex on demand and the inferiority of women. These ideas are of the same cloth, however. Whether coming from the pulpit or from social media culture, ideas like these produce significant harm primarily, but not exclusively, to women.

Ending intimate violence is a shared Christian responsibility. We need to begin by correcting and eradicating theologies that perpetuate harm, including those that valorize involuntary self-sacrifice and suffering, sexism, and gender dominance, as well as those that negate personal and communal responsibility by claiming "sinful behavior is inevitable." Instead, Christian sexual ethics can focus on healthy relationship values and spiritualities of resistance that value erotic attunement. A key aspect of such education is digital literacy about what sexual harassment and intimate violence look like online.

Digital Sexual Harassment and Intimate Violence

Violence can and does happen through all the technologies discussed in this book. But is it the nature of the technology that is the main problem, or that of the human usage? Other than surveillance devices and software, which people will say are intended for safety, everything else appears to be built for relational, educational, and pleasurable intentions and not violent ones. Yet, as described above, these same affordances or benefits of a technology can be exploited by users for harmful or abusive purposes. These online behaviors morally deform those participating. As theologians Rob Rhea and Rick Langer note, "Actions taken in an online context have a direct connection to and are the responsibility of the person acting. These actions are extensions of relatedness in the same way that extending one's hand or body is an initiative toward relatedness."[15] In addition to the moral malformation of users, there is collateral moral damage done when individuals produce and distribute sexually violent content. Most platforms employ workforces, many located in the global south, to review online content and weed out illegal, harmful, and violent materials.[16] Often, artificial intelligence is used to flag this content, but then humans need to assess whether the breast in the picture is a breastfeeding mother or a naked, underage user. This is a relatively benign example; platform content moderators see hideously violent acts repeatedly in a single day's work.[17] When considering the implications of sexual violence and digital technology, we need to be attentive to the larger ecosystem in which these technologies function.

Based on the various technologies discussed throughout this book, this section will focus on actions that overlap technologies, such as online harassment via online dating or communication applications, cyberstalking and cheating apps, revenge porn and use of deepfakes, and finally the impact of sexual violence perpetrated in online environments. Unlike other chapters, I will not spend time discussing how one uses these digital technologies, because that would reinforce violent or abusive uses of technology. Rather, I will describe the legal and technological ways to counter these behaviors.

Consider the following accounts of victim-survivors from journalists trying to expose the pervasiveness and complexity of online sexual harassment and violence:

> My ex-boyfriend, who is Italian and living in the U.S., has turned into a very malevolent cyberstalker and cyberharasser. I am a British citizen, living in Hong Kong. This mix of nationalities and countries of residence makes it difficult to deal with the situation: Which jurisdiction is relevant? Who will take interest in this and help me find a solution? The websites used by people like my ex allow anybody to post anything—material which is always slanderous—without ever checking facts and in the name of "freedom of expression." It is then nigh on impossible to have these posts removed without going through lawyers, at vast expense. They are, in effect, extortion sites. Cyberstalking and cyberharassment are becoming more endemic and increasingly cross-border. International victims seem to fall between the cracks of national jurisdictions and national regulatory authorities. As a result of these posts, both my personal and professional life have been seriously impacted—and there is no one I can turn to. What should I do? —Diane[18]

Jay's wife, Ann, was supposed to be out of town on business. It was a Tuesday evening in August 2013, and Jay, a 36-year-old IT manager, was at home in Indiana with their 5-year-old daughter and 9-year-old son when he made a jarring discovery. Their daughter had misplaced her iPad, so Jay used the app Find My iPhone to search for it. The app found the missing tablet right away, but it also located all the other devices on the family's plan. What was Ann's phone doing at a hotel five miles from their home?

His suspicions raised, Jay, who knew Ann's passwords, read through her e-mails and Facebook messages. . . . He didn't find anything incriminating, but neither could he imagine a good reason for Ann to be at that hotel. . . . Jay spent a few days researching surveillance tools before buying a program called Dr. Fone, which enabled him to remotely recover text messages from Ann's phone. Late one night, he downloaded her texts onto his work laptop. He spent the next day reading through them at the office. Turns out, his wife had become involved with a co-worker. There were thousands of text messages between them, many X-rated—an excruciatingly detailed record of Ann's betrayal laid out on Jay's computer screen. "I could literally watch her affair progress," Jay told me, "and that in itself was painful."[19]

It was late on a school night, so Jennifer's kids were already asleep when she got a phone call from a friend of her 15-year-old daughter, Jasmine. "Jasmine is on a Web page and she's naked." Jennifer woke Jasmine, and throughout the night, the two of them kept getting texts from Jasmine's friends with screenshots of the Instagram account. It looked like a porn site—shot after shot of naked girls—only these were real teens, not grown women in pigtails. Jennifer recognized some of them from Jasmine's high school. And there, in the first row, was her daughter, "just standing there, with her arms down by her sides," Jennifer told me. "There were all these girls with their butts cocked, making pouty lips, pushing their boobs up, doing porny shots, and you're thinking, Where did they pick this up? And then there was Jasmine in a fuzzy picture looking awkward." . . . You couldn't easily identify her, because the picture was pretty dark, but the connection had been made anyway. "OMG no fing way that's Jasmine," someone had commented under her picture. "Down lo ho," someone else answered, meaning one who flies under the radar, because Jasmine was a straight-A student who played sports and worked and volunteered and was generally a "goody-goody two shoes," her mom said. She had long, silky hair and doe eyes and a sweet face that seemed destined for a Girl Scouts pamphlet, not an Instagram account where girls were called out as hos or thots (thot stands for "that ho over there").[20]

These scenarios are real and pervasive. Each of these stories use pseudonyms to protect the identity of the journalists' sources. From an undesired dick pic to reposting of shared photos and videos, online sexual harassment and stalking have increased. A recent Pew Research study notes that "growing shares of Americans report experiencing more severe forms of harassment, which encompasses physical threats, stalking, sexual harassment and sustained harassment."[21] This could be in part due to an increased use during the COVID pandemic of social media, the main vehicle for such abuse, and online communication. Regardless of the cause of the increase, it is a fact that online spaces include experiences of sexual harassment and abuse, in addition to other forms of harassment. Unfortunately, demographic factors for increased abuse—such as gender, race, sexual orientation, and age—remain the same online as offline. For example, people are more likely to experience online sexual harassment if they are female-identified, are Black or Hispanic, and are young adults—as opposed to white, heterosexual, middle-aged men.

Lesbian, gay, and bisexual individuals are most likely to experience any form of harassment online, not just sexual. In order to combat practices of unwanted surveillance and exploitation, one needs to know current technology and privacy regulations and how to reach out to site and platform hosts to report abuse.[22]

Online Dating or Communication Apps

In chapters 1 and 2, I noted how sending sexually explicit material through direct messaging can enhance or harm a relationship. The key difference is consent. Jasmine, in the last vignette above, received repeated pleas from an older boy at school to send him a nude photo. She and many other young people incorporated sexting (by text, image, and video) into their relationships. While she didn't feel comfortable sending it—a clear red flag that signals coercion—she convinced herself it was a way to show she liked the boy and hopefully gain his affection. Sadly, it is not unusual for a personal, nude photo to be shared beyond the intended recipient. Yet, the outsized examples of "sexting rings"—where a small group of young men post and rate all the girls in their school or all the young women in the campus sororities—are less common, though they gain significant media attention. Many municipalities have had to create different legal structures to respond to self-produced pornography featuring children (anyone under eighteen). In the past, these cases would be dealt with as child pornography, requiring prosecution of the teen girl who produced and sent the picture of herself. As sexting has become a common sexual behavior for adults and teens, laws are slowly adjusting, and so are school conduct policies.

In other circumstances, the sexually explicit material is completely unsolicited. YouGov, an international research data and analytics group out of the UK, found that four in ten women aged 18–36 had received an unsolicited photograph of a penis.[23] One might wonder who would send a nude photograph without being asked. In "I'll Show You Mine So You'll Show Me Yours," researchers report motivations for sending "dick pics" range from wanting a photo in return to hoping to elicit sexual excitement. The senders of these images also exhibit higher levels of narcissism and ambivalent or hostile sexism.[24] There is a clear gender double standard at work in harassment related to self-produced sexually explicit photos and messages.

Digital technology makes it easy to receive (even when unwanted), share, or reproduce personal communication. The ease of the digital technology provides an avenue for harassment and harm that previously did not exist. In some cases criminal prosecution is an option. Yet legal remedies quickly fall behind digital technological innovation. Not to mention that those with the fewest resources tend to have the least access to and trust in policing and legal responses. More likely, a user needs to block a sender or report their behavior to site administrators to have them blocked.

Cyberstalking

Similarly, cyberstalking allows for a new scale of behavior previously not possible. In the past one might ask around about a person or follow and observe them at a party (before or after) dating. But now one can practically do a private investigation background search on a potential date prior to meeting them and keep track of their whereabouts or communications with savvy use of digital technologies. One can find out a lot about a person from an internet search or following the trail of networked relationships across social media platforms. Honestly, a little online investigation of a person before getting into a personal relationship might be a good thing. However, delving into every aspect of someone's online record can be invasive and ultimately reveal information that is out of context or no longer true.

Digging into publicly available information, like internet search results, does not legally qualify as cyberstalking. On the other hand, some forms of stalking use basic software enabled features like Find My iPhone, described in the second scenario above, or Snapchat's location map. Cyberstalking, like most forms of intimate violence, is overwhelmingly done by someone the victim-survivor knows and includes a range of behaviors, but most basically it requires the use of digital technology to follow, control, and intimidate.[25] Some might respond with the advice to "just get off the internet," but that is a victim-blaming approach. Beyond that, it is almost impossible to disconnect and still live in today's society. A mobile phone is a necessity, as are email, banking, and credit cards—all of which leave a digital trail even without social media use. In addition, online harassment rarely stays online. It can affect job prospects and other relationships. And fear and intimidation cause offline physical and emotional issues for victim-survivors, not to mention the real and sometimes realized fear of a perpetrator showing up in person to cause harm and violence. Cyberstalking can happen during and after a relationship. One might be living with the very person who uses cyberstalking as a form of control and abuse.[26]

In the case of Diane, the cyberstalking started after her breakup. Her ex-boyfriend exploited the fact that platforms do not police the content posted on their sites; so he could post harmful lies if he chose. Some pertinent regulations are changing on major social media sites, and sometimes the host of a platform will intervene if a user files a complaint. However, the scale and reach of the information are often so wide and far that it is difficult to repair reputational damage once it has occurred. Also, like most offline stalking and harassment laws, legal recourse is based either on the country in which the crime took place or the place it originated. This can leave victim-survivors at a profound disadvantage since the internet does not function with the same geographic or governmental limitations as legal responses require. Thankfully, there are advocacy groups and legal partners working to remedy the inadequate patchwork legal and policing methods needed to combat cyberstalking.[27]

Tracking someone online may not always be viewed as a morally harmful pursuit. For example, some readers will side with Jay's use of GPS and software monitoring of his wife's correspondence to discover her cheating. And later in the story, we learn that Jay and Ann decide to work through the infidelity and rebuild their relationship. One key aspect of rebuilding trust is the continued use of the technologies by Jay to allay his fears that Ann might lie about her whereabouts or actions. The use of these technologies in this circumstance builds trust, through concrete checks on Ann's honesty. But I'm not convinced this is *trust* if it relies on a technology to confirm facts. In other cases, those who cheat use equal amounts of apps to cover their trail as their partner might use to discover their location or communications. Those behaviors create relationships marked by distrust, dishonesty, jealousy, and lack of shared communication.

Revenge Porn and Use of Deepfakes

Revenge porn—posting of sexually explicit content such as direct messages, pictures, and videos without one's consent—is commonly thought to follow a relationship break-up. However, it can also be used to coerce someone to stay in a relationship or scorn a person for turning down a relationship or not engaging in specific sexual behaviors. As has already become clear, women are more often the targets of this exploitation and harassment. The material can be content that was willingly shared with a partner, recorded or taken without permission, or created as a deepfake by combining a public piece of pornography with a video or photo of the victim-survivor. In Jasmine's situation, the reposting of her photo was not revenge porn, but similar coercive tactics related to the promise of a relationship were used to encourage her to send the photo. It's possible that Diane's ex has posted videos or photos of her or even created fakes; she noted that the material is slanderous and had been posted without checking facts.

In cases when someone consensually shares material with another person, the only recourse one has to the privacy of that material is to argue ownership of it, as with a copyright. However, if another person is in the video or the material was created on someone else's device, for example the partner's phone, then ownership gets more complicated. Lack of consent for production of the material also assists a victim-survivor in having the material removed, though that can be difficult to prove. The complexities of ownership of digital material often mean it is shared and reproduced without penalty. Those who post revenge porn count on this complexity as a form of protection. On top of that, lax legal publication standards do not hold platforms accountable for content, and thus there is very little liability for defamation claims.[28]

Deepfakes—created through a process of deep machine learning—usually combine a person's head with another person's body and can realistically simulate movement and even speech. Some deepfakes are so well produced that even another deep learning program cannot identify the "tells" that mark them as fakes.[29] Most deepfakes are pornographic, and the people depicted have rarely given their consent. The technology is within reach of an everyday computer user and thus could thoroughly change the way we perceive the truthfulness of information, especially as it relates to the way deepfakes have been used against political figures.[30] As deepfakes show up on social media platforms, having migrated from other sites that have been shut down or that have banned them, it has become clearer to legal advocates that new laws are needed, such as a requirement for explicit consent from all parties depicted in a photo or video and holding platforms responsible for vetting content.[31] In any case, legal remedies will not keep pace with technology, nor do they constitute an avenue available to all. Instead, theologian Clifford Anderson wonders about the role of empathy in response to deepfakes: "Believing in spite of appearances. Disbelieving despite the evidence. Seeking for signs or 'tells' contradicting what otherwise appears genuine. Keeping faith in the other, but not falling prey to false messiahs. Are these traits of empathy in an age of deepfakes?"[32] Certainly, users will need to lean into values that counter the motivations of jealousy, anger, coercion, and so on behind deepfakes.

Sexual Violence Perpetrated Online

The differences among sexual harassment, violence, or abuse are fine lines drawn for legal purposes. Sexual violence perpetrated online could include anything from livestreamed rape videos to sexual assault of an avatar in virtual reality. In one case, people are interacting in person and sharing content. But it could equally be the case that no other person aside from the user is involved and the "other" is a computer-generated image. Some advocates also include sex trafficking in this section because the transaction of selling another human being for sex originates online via social media, a message board, or a video game.[33] All of these examples are nonconsensual acts and not produced as fictional fantasies.

Some pornography depicts consensually produced rape scenes or sexual assault performances. While some argue that such content should be banned, it is legally and morally different than nonconsensual acts. This also raises the question of sexual practices online that intentionally include violence like bondage and discipline, domination and submission, and sadomasochism or BDSM. Again, some critics think any inclusion of violence, even when consensual, in the expression of sexual behaviors leads to moral malformation of sexuality. Others, however, argue that those who engage in BDSM online or offline employ robust consent agreements, use safe words to automatically stop a behavior, and put the care and pleasure of another first in ways that most sexual encounters do not. BDSM, with these stipulations, thus requires trust, honesty, and clear communication.[34] Of course, what one sees in online pornography may not include all of the negotiations that happen before or during engagement in BDSM. Presumably, however, if one were to engage in BDMS, say in virtual reality, these conversations would take place. Thus, not only is consent a key feature in determining the moral and legal aspects of sexual violence, truthfulness of related content depictions are necessary.

Nonconsensual sex acts constitute assault and violence. However, when the sexual behavior happens online to a person's avatar or image, how is it categorized? And does it only have moral importance if there is another human behind the avatar? As we will discuss in chapters 4 and 5, even in online spaces users are embodied and they affect and are affected by their actions in relationship to others, whether those be humans, avatars, or robots. In response to something like sexual assault of an avatar or robot, we might create different legal repercussions for an abuser. This is primarily based on a desire to avoid judicial precedent that makes a robot or avatar legally equal to a human person. However, from a moral standpoint, the use of sexual violence to exercise power over another person is a harm not only inflicted on the other, but on oneself as well. If a person experiences nonconsensual sexual violence while embodied as an avatar, that person still feels the violence in an embodied and affective way that could traumatize them. If a user exploits a backdoor in a game or VR environment to perpetrate sexual violence on another— whether a person or an avatar or robot because the user imagines it as a "human"—the perpetrator's moral character is damaged by intentionally dehumanizing another.

Similarly, sex trafficking—the purchase and sale of another human being for sexual behaviors—is sexual violence. The entitlement created by sexual, racial, and economic oppressions leads some to believe they can use humans as objects. As Ellison writes, "In a white racist, class-stratified, and patriarchal social order, some groups have historically claimed entitlement to exercise power as control over other groups while those with power are obligated, by force if necessary, to show deference and serve the interests of those 'above them.'"[35] No one is another person's property. No one has a right to another person's body, even if the two people are married, even if they have engaged in sexual behaviors in the past, even if one owes the other money, even if they are projected online as an avatar, and especially not if they are a child.

Sexual harassment, assault, and abuse are wrong. Legal and social advocates work tirelessly to end sexual violence through better legislation, advocacy and outreach programs, and education that includes digital literacy. In faith communities, we need to recognize how rampant it is online and offline, to stop theologically contributing to the problem, and to seek ways to help with healing and prevention.

Sexual Ethics and Spiritualities of Resistance

As is clear in this chapter and in upcoming chapters, many digital technologies facilitate holistic sexuality education and sexual empowerment for women, LGBTQ persons, and people who are differently abled, to name a few. On the other hand, routine concerns over digital technologies and sexual health focus on heterosexual men who are mostly white and young. These overwhelmingly constitute the same demographic that abuses pornography, sends unsolicited sexual messages, nonconsensually shares past sexual material, and harms and harasses others because of their gender or sexual orientation. Andrew J. Bauman, a Christian counselor, dedicates his ministry "to healing and restoring brokenness within male sexuality."[36] He does this by helping heterosexual men unlearn harmful, sexist theologies and explore healthy sexual expression and gender equity. He does this through a body-affirming and sexuality-positive Christian approach that pushes beyond masculine socialization.

Sexual ethics based in Christian values guided by the love commandment and personal accountability for erotic attunement needs to ground Christian theology, teachings, and expectations for sexual health and sexual relationships. The legacy of heterosexist Christian sexual ethics feeds into a dismissal of sexualized violence. Instead, Christian sexual ethics should expect and nurture values of honesty, consent, and respect in all relationships, but especially in sexual relationships. As Ellison notes, "without mutual respect, reciprocal vulnerability, and power shared consistently and fairly between intimates, intimacy is compromised."[37] In order to be mutual and reciprocal, partners need to be attuned to their sexual desires and feelings, and to restrain them if expressing or acting on them would harm the other person. This type of attunement or discernment comes through education and practice in relationships. It also requires a larger scale social and theological shift away from heterosexist and patriarchal theologies that perpetuate domination of people based on gender, sexuality, race, and so on.

Unfortunately, many of us respond to issues of sexual violence and harassment as bystanders who choose not to intervene; we do not prioritize policies and practices that hold perpetrators accountable. Feminist Christian theologian Elisabeth Vasko believes the position of the bystander requires theological exploration to move Christians from passive complicity to spiritual and physical resistance. She writes that "we are born into a world where patterns of human violation—racism, violence against women, heterosexism, imperialism and classism—are already set in motion. Yet, regardless of intent, we are responsible. Injustice that is overlooked or ignored is dangerous. . . . The habit of ignoring suffering bodies is difficult to break."[38] Throughout her work in *Beyond Apathy: A Theology for Bystanders*, Vasko details how and where we can break the habit and begin to condemn and respond to suffering and violence in our world, as did Jesus.

The FaithTrust Institute, an organization that has for decades worked toward the prevention and eradication of sexual harassment and abuse in faith communities, suggests a three-pronged approach: (1) recognition, (2) prevention, and (3) intervention.[39] Recognition means being better informed about the prevalence and impact of sexual harassment, violence, and abuse, especially in its online forms. As for prevention, many faith communities in the past twenty years have put child sexual abuse prevention policies and practices into place. Most denominations provide resources for these policies as well as requiring various forms of sexual misconduct and harassment prevention training for clergy and some staff and volunteers. But do these policies include education related to digital technology or digital means of harassment and abuse? In addition, a commitment to prevention also requires that all Christian communities examine how their theologies and their lived practices may contribute to or be used to justify sexual violence and abuse, to protect abusers, and to silence victim-survivors. Shared theological commitments to equality, compassion, and justice can serve as a resource for healing and wholeness as well as a moral charge to end sexual and domestic violence. Last, when an incident does occur, policies and practices should already be in place so that those responding can be clear about their roles and responsibilities, including legal requirements for reporting and intervention steps to secure the safety of victim-survivors.

In faith communities, spiritualities of resistance to all forms of oppression strengthen the whole community. As West notes, "Churches can ... play a unique role in helping to withdraw their own and other communal supports for the abuse and violence. Developing an antiracist approach can aid in producing the Christian moral imagination needed to awaken from a tolerance of everyday traumatic consequences for those who experience intimate violence and abuse or the ongoing threat of it, especially the most socially marginal community members."[40] No faith community is immune to sexual harassment and violence, though these may take different social forms and impacts based on the community's context. Christian communities can and should be places that lead in efforts toward healing, education, and prevention.

Everyday Digital and Sexual Ethics for Ending Violence

- Christian theology can be used to perpetuate harm and violence.

  Do you ascribe to theologies that suggest that all forms of suffering—voluntary and involuntary—are good? Are you part of a religious community that prioritizes heterosexual men's leadership but does not hold them to standards of honesty, mutuality, integrity, and respect? What do you learn about masculinity from Jesus's example of caring for others, treating others equally, and pushing back against the social conventions of his time? What kinds of sacrifices for another person make a relationship stronger rather than support domination or control?

- Faith communities can be informed advocates.

  Your faith community's advocacy committee could reach out to local leaders to learn about the laws in your area on cyberbullying, stalking, and sexual harassment. Hold an educational session for church members. What do the youth, parents, and other adults in your community need to know? What current legislation or policies need shared advocacy or revision?

- Digital privacy practices require attention and cultivation.

  Have you done a privacy checkup on your social media accounts or web browser? Is there any need to have social media accounts set to public or allow anyone to direct message you?

Do you know how to block another user or messenger on the applications you use? What information do you share with others and through what applications; for example, are they encrypted services?

- A Christian response to sexual and intimate violence requires spiritualities of resistance.

  How does your faith inform the way you understand gender equality and sexual inclusion? What theological beliefs or Scripture stories support a spiritual resistance to sexual and intimate violence? What are the local survivor resources in your area? What education is needed to identify and respond to online and offline patterns of abuse and harassment? How can you become an upstander, a person of faith publicly seeking an end to sexual violence?

4

Where Two or Three Are Gathered

SEX AND INTIMACY IN VIRTUAL REALITY

H AVE YOU EVER WATCHED A MOVIE or read a book and felt affected by the story—physically reacting to the pain, joy, or
pleasure and creating a relationship with the characters? Could
you relate to them "like real people" and maybe even learn from
their mistakes and successes? Did you cry and mourn when one
of them died, or feel jubilant when they were finally reunited
with a loved one? Or maybe you had a crush on one of the characters, fantasizing about what it would be like to be with them?
Great stories create new worlds and spark imagination as they
invite us to bring ourselves into a relationship with the characters. Virtual reality (VR) allows users to physically enter a story,
a new world. *Physically*—what does that mean? Some hear the
word "virtual" and replace it with "fake." Virtual reality is an
alternative space to offline experiences. Both require participation of bodies to react to the material or physical aspects of these
spaces. There are a variety of ways to engage in virtual or augmented realities, from control of an avatar in the popular Second
Life community game[1] to a fully immersive VR headset like the
popular Oculus device.[2] Each form of engagement requires use
of senses, cognitive engagement, and responses to the other(s) in

the experience. Sexuality as an embodied experience is not left behind when one enters virtual reality.

There are a variety of ways to engage sexual experiences in VR. "Digisexuality" is a general term used to describe digital sexuality, including all the topics covered in this book. More specifically, "digisexual" is used to describe a sexual orientation and practice where one's sexuality is oriented toward and fulfilled through digital technologies. Researchers Neil McArthur and Markie Twist write, "We use the term digisexuals for people whose sexual identity is shaped by what we call second-wave sexual technologies. These technologies are defined by their ability to offer sexual experiences that are intense, immersive and do not depend on a human partner."[3] In some cases, these individuals consider their sexual orientation or attraction and desire to be directed toward and fulfilled by engagement with digital technology—meaning they no longer seek sexual behaviors with or desire sexual interaction with other humans. This is a subset of those who use various extensions related to software-enabled artificial intelligence (AI) devices and virtual reality for sexual pleasure.

Many others have what we would call a hybrid approach and may use VR- or AI-enabled digital devices to enhance offline human relationships. For example, a long-distance couple might meet up in a VR game or use videoconferencing and "teledildonic" devices controlled by their partner for remote pleasuring. Teledildonic devices are hardware devices like sheaths or dildos that can be controlled directly by software applications, usually on a user's phone or tablet, or that can be connected to software in a VR game. These devices rely on haptic technology that stimulates the user by mimicking bodily touch and sensation. There are also VR sexual experiences that allow a user to be present in a three-dimensionally recorded pornography video. Rather than passively viewing, the user feels like they are present and participating. Other video and narrative games, which are expanding to include LGBTQ characters, allow users to play through their avatars.[4] Virtual reality engagement thus covers a spectrum of devices and software. In some spaces, a user might be a remote-controlled avatar player who can meet others and engage in sexual behaviors alone or with other avatars. Or a user could have first- or third-person immersive participation in a VR scene that may or may not include teledildonics controlled by a partner or an AI.

The first introduction to these technological advances can feel like a sci-fi future gone awry—a dystopic, post–climate crisis world described in novels like *Ready Player One* and *Ready Player Two*, where almost all human interactions take place in the virtual reality universe called OASIS.[5] And yet, characters can experience almost any sexual relationship or behavior vicariously through VR, which ends up expanding sexual inclusion and diversity as first-person virtual experiences reduce individual biases and judgment. Additionally, as Edward Evans Bailey notes, there are advantages to these forms of virtual intimacy for individuals who experience physical disabilities that limit their movement, or for those who have social anxiety or more extreme mental health issues, or simply for those who wish to avoid sexually transmitted disease and pregnancy. For some this might be their only option for sexual intimacy and behaviors—or at least safer and educational ones. He also suggests this technology might have wider uses, like for "the skewed gender ratios in some countries, which may leave millions of [heterosexual] men unable to find a (human) mate. Or you could think of VR simply adding a new immersive element to existing virtual text, phone, and webcam sex, all of which are nothing new."[6] These sexual experiences thus allow users to connect with other humans and AI in ways that provide wider options for sexual exploration and intimacy. They may also skew toward male heterosexual users looking for the ideal female and allow for exploitative behavior similar to that connected with some forms of pornography consumption. In that case, the technology itself may not be deemed unethical; rather, it's the intentions and behaviors of users that can exhibit harm.

Sexual intimacy and behaviors enabled through software and hardware, or with virtual reality, raise the question of whether they deepen an incarnational experience of sexuality and body or disconnect us from it. Scriptural traditions refer to sexual intercourse as "knowing someone"—giving us a richer conception of the role emotions and bodies play in sexual learning and connection. "Knowing" also reinforces the idea that sexual behaviors teach people about themselves, not only about others. Additionally, this embodied form of knowing resists the stereotypical and limited understanding of reason or rationality as disembodied. Historically, the defining feature of what it means to be human has been dependent on a person's capacity for reason or higher thinking. That standard raises questions about which criteria we use to define humans. This is a critical question if one of the objections to sexual behaviors in a virtual space is engagement with "nonhumans."

In other words, is it really sexual intercourse if someone with a headset sees and feels themselves penetrating a computer-generated image, whether or not that image is connected to another user, known or unknown, on the other end? Is there intimacy or embodied knowledge in the encounter? The use of avatars or other bodily representations "signifies the importance of some form of a body to fully inhabit the online world."[7] Perhaps sexual behaviors in virtual reality will promote social goods, such as decreasing sexually transmitted diseases and unintended pregnancies, allowing users to explore different genders or sexual orientations, or offering intimate options for those with a range of disabilities, or providing practice at relationships, including sexual behaviors, to gain facility and skill. Sexual expression in virtual reality is embodied, but also not bound by the limits of one's own skin.

Human Embodiment and Incarnation

The Christian tradition has had a complex relationship with the body and what it means to be human. Humans originate from dirt or earth infused with the breath of God, as narrated in the creation story. The synapses and sinews, muscle fibers and bones come together as earthly materiality formed through a spiritual process. In this way, humans are embodied spirits, or spiritual embodiments, not to be disentangled. Embodiment is connected to nature, and yet humans are thought of as a special class of creation and often considered better than nature (nonhuman animals, plants, water, rocks, and so on). Over the centuries, scientists have debunked the myth that humans are the only species with cognitive abilities like language, memory, and future planning. Obviously, humans are distinct, but are we distinct only for the ability to reason or cognition? Is skin merely a container for materiality, a prosthesis of sorts? In the search for a carnal theology, Bonnie J. Miller-McLemore notes, "At some visceral, pre-rational level, the body learns and knows, and thought follows."[8] The body contains knowledge in the sense that the brain is a bodily organ and that we learn through sensory perception, synapses that are material interactions, not disembodied free flowing thoughts. We don't only have bodies; we are bodies.

Different from the reliance on reason or cognition as the defining feature of humanness, sometimes embodiment is reduced to the material ways humans are defined by sex, gender expression, race, and physical abilities. That is to say, embodiment is often used as code for the ways human identities are culturally constituted and understood. Identifiers like race or gender can be used by society and religious communities to include or exclude certain people.[9] Cultural and religious messages shape how we understand and value bodies and embodiment based on external markers. These messages directly affect the formation of people's sexualities and relationships to their bodies, often reducing sexuality to body image or racial and gender stereotypes.

In seeking to define the connection between humanness, embodiment, and sexuality, Christians often look to the creation stories in Genesis but fail to consider the incarnation. The truth of the incarnation suggests that God is met by us and for us in the body. I suggest that we take a closer look at the meaning of an incarnate Christ to find insights and relevant theological meanings for embodiedness. James Nelson, who in the 1970s theologized about embodiment and sexuality, suggests bodily concreteness becomes a context for religious reflection, our particularity becomes not a limitation but an "incarnational gift."[10] Christ, as fully divine and fully human, reinforces the notion of relatedness and self-awareness as part of physical presence, strengthening an awareness of God in relation and in our relations. The positive attributes attached to this understanding of bodiliness run counter to subjugating the body in the service of reason as if the mind were not part of the body. We can affirm that humans know through our bodies.

We are spiritually embodied beings in ways that rely on nature and share features with other animals but also in ways that are distinct. Theological traditions that want to create stark divisions between "nature" and humanity usually do so as a way to argue for human superiority over and exploitation of nature and non-human animals, rather than for relatedness among and through God's creation.[11] To this relational view of humanness, we now need to add the digital dimension.[12] Humans use technologies to extend the limits of our bodies. When interacting with technologies, we are not always cognizant of how we are being reshaped in the process. For example, when we use a shovel to dig a hole, we often see this as a tool under our bodily control, and yet the user becomes a much more efficient digger when using a shovel, and probably later in the day their hand is aching and blistered from the reshaping of muscles and friction. The digger is changed physically and mentally in the process. These types of embodied changes are exponential with digital technology and often more hidden from our direct perception. For some, the breakdown of the human-machine distinction is a recent event because of the proximity of integration of technology into our bodies (like constant connection with smartphones, earbuds, pacemakers, or mRNA treatments, all of which bridge tenuous and fabricated divides between nature, machines, and humanity).[13] Others believe we have been transhuman for centuries—since learning to use tools, wear glasses, or since coming to understand the synaptic chemistry of the brain.[14] As the fabrication of the "man-made" distinctions between humanity, nature, and machine become more recognizable, questions arise about qualities once considered unique to humans (like cognition, decision-making, emotions, and so on).[15]

The advance of digital technologies provides a novel way to experience ourselves. Theologian Kutter Callaway suggests that "the interfaces we employ to connect with our digital environments provide us with both a model and a means for being and becoming more fully human."[16] Callaway argues for a relational understanding of humans that sees bodies as embedded in digital environments. What does that mean? Humans regularly augment or add to their bodies using technologies like those already mentioned: glasses or watches. Some are even more integrated with the body like corrective lens surgery, pacemakers, hearing devices, or synthetic ligaments or bones. Most often these devices aid communication or enhance human abilities. Humans incorporate these technologies into their bodily schema. We do so even when they are not embedded in our bodies. Think about the phantom feel of your smartphone in your back pocket and the jolt of realization when it's missing, or the weight of earbuds in your ears even when they are removed after hours on video calls. These devices leave a sensory imprint. Callaway suggests these experiences can help us understand a kind of theological anthropology that shows us that humans—as embodied, digital, spiritual beings—are "body-centric" not "body-bound."[17] In other words, the body, understood incarnationally, is central to being human, but our skin is not the boundary at which incarnation ends. It is the conduit or interface for relationship, especially in the digital realm.

Some scholars interpret digital-human interaction as moving humans away from God's original blueprint for humans, even calling it "excarnation," a leaving behind of our spiritual embodiment. This assumes that because virtual reality or other digital technologies are supplemental to the body, they somehow remove users from their bodies.[18] Callaway argues the exact opposite, and I would agree. Digital technologies ground us in our bodies in new ways that create new possibilities. They allow us to extend ourselves into new spaces and interactions through networks that are material extensions of computing (even when we don't see the material aspects of software and wireless waves). We have a theological metaphor to describe this reality—the body of Christ. What if the body of Christ isn't only a metaphor? And here Callaway cites Matthew 18:20: "In other words, the extension of our bodies (through digital and physical means) into the body of Christ creates the necessary conditions for Christ to be present—real, actual, effective—in the first place."[19] The extended body, or body-centric but not body-bound, experience of personhood exemplifies incarnational theology. As we make Christ present in our embodied interface with digital extensions, we create new meanings through relationships with self and others.

The interpretation of body-centric incarnational theology resonates with the broader definition of sexuality I have used throughout this book. Intimacy and sensuality become primary ways of knowing and expressing one's sexuality. This aligns with scriptural uses of "knowing" as it relates to sexual encounters that value embodiment, not to disembodied reason. To conclude, being human isn't only having a body; it is being a body. In and through the body, we learn, interact, and extend ourselves into new relationships and networks that bring the communal body of Christ into being.

Sexuality in Virtual Reality

The possible forms of sexual behavior—not to mention types of sexual expression—in virtual reality are wide and diverse. For this portion of the chapter, we will focus on sexual encounters in virtual reality that include avatars and are accessed through immersive-reality hardware and software. This includes a level of extended embodiment that provides first-person interaction with others, human users or AI, through neural networks along a scale of software sophistication that fosters immersive bodily response. The Netflix series *Black Mirror* has had two different episodes exploring sexual relationships in virtual reality. The "San Junipero" episode of *Black Mirror* plays with the concept of a virtual reality afterlife for the elderly, allowing them to have relationships as twenty-year-olds.[20] Another episode, "Striking Vipers," offers an example of how VR gaming creates opportunities to explore sexual behaviors with other gamers.[21] While we may not yet experience neural simulated realities or the digital existence dramatized in these narratives, we experience other possibilities in current virtual reality technologies.[22] "San Junipero" is like the next generation of the 3D Second Life community that finally got the upgrade to immersive VR by tapping into the user's neural network.[23] "Striking Vipers" uses similar neural network technology but is experienced as a digital video game in which one "plays" rather than lives, more like the current Oculus VR games.

"San Junipero" depicts a simulated reality of the same name created for the elderly to visit and eventually choose as a place to live their afterlife uploaded to the cloud. In their twentysomething simulated bodies, residents can visit distinct decades, returning to the time of their youth or trying out new time periods.[24] Yorkie and Kelly, the main characters, meet in a 1980s bar scene. Yorkie, a white gangly young adult who doesn't drink or dance, is visibly uncomfortable and stands out. Kelly, a self-confident, stylish Black young adult, is trying to elude Wes, an overbearing white dude who begs Kelly to have sex with him again. Having just met, Yorkie plays the role of an old friend at Kelly's request and helps get rid of Wes. Kelly discovers this is Yorkie's first time to San Junipero and wants to introduce her to the benefits of unencumbered fun. Yorkie, in her discomfort, leaves the bar. The episode initially reinforces sexual stereotypes related to Kelly as a sexually experienced, pleasure-seeking Black woman and Yorkie as a white, virginal, rule-follower. In their exchange, we find out that Yorkie identifies as gay but has never had a relationship or explored her sexual orientation. Kelly volunteers to help her. This leads to a dramatic, young adult back-and-forth relationship. One night in bed, Kelly suggests they should see each other in real life. Yorkie first dismisses this idea but eventually gives Kelly her geographic location.

Offline, Kelly is an elderly Black woman with cancer whose husband and child have preceded her in death. She goes to visit Yorkie, and she meets Greg, the man Yorkie says she is going to marry. Kelly learns that Yorkie has been a quadriplegic since she was twenty-one, when she crashed her car after coming out to her parents, who rejected her. Yorkie wants to be euthanized so she can permanently live in San Junipero, but the law requires a family member, a lawyer, and a medical professional to consent. Her family continues to punish her by refusing to sign. Greg is her nurse and has agreed to marry Yorkie so he can sign as her family member. Now understanding Yorkie's full story, Kelly requests five minutes to see Yorkie in San Junipero, where she asks whether she can marry her instead of Greg. With this complete, Yorkie is euthanized and uploaded to live in San Junipero for her afterlife.

When Kelly reunites with Yorkie in San Junipero, they argue over Kelly's commitment to be buried with her family when she dies, not uploaded to the cloud in San Junipero. Regardless of Yorkie's pleas that "this is real" (motioning to her surroundings) and that "this is real" (affectionately touching Kelly), we are led to believe Kelly will not return. When Kelly finally decides to be euthanized, the scene cuts to her body being buried next to her previously deceased family members in the graveyard, and then her data or consciousness is placed next to Yorkie's in what looks like a mini-robotic mausoleum. In the final scene, Yorkie picks up Kelly at their San Junipero beach house and they (presumably) live happily ever after in death.

San Junipero, as a virtual reality, allows users to play with previously fixed notions of sexual identity, boundaries of age, and racial purity in sexual relationships.[25] The technology allows these two users to be in a new and ongoing sexual relationship that we might consider akin to offline experiences. In San Junipero, users discover not only the fullness of their sexual orientations, but they also love one another. VR relationships can enhance or turn into offline relationships.[26] In fact, that's the whole purpose of San Junipero, to give the elderly a way to fully engage in relationships free of limitations imposed by aging. The episode highlights the need for intimacy, sensuality, and love throughout a lifespan. VR is an avenue to extend bodies into new spaces and continue to experience these important aspects of being human.

"Striking Vipers" offers the viewer a similar narrative of intimacy and sexual exploration, but it upsets more traditional understandings of coupling. This episode features an all-Black main cast. In it, best friends Danny and Karl reunite in their late thirties at Danny's birthday party. Karl gifts Danny the latest version of virtual reality hardware for a video game, called Striking Vipers, they used to play in their twenties when Karl, Danny, and Danny's then-girlfriend and now-wife all lived together. The newest version of Striking Vipers is a virtual reality martial arts game operated through neural network technology. Users feel the pain of the fight, and we later learn they can also engage in multisensory, embodied sexual behaviors.

During the birthday party, Karl and Danny catch up and talk about their lives. Karl jokes with Danny about the hot, younger women he hooks up with using a meetup application. Danny reflects on his out-of-shape dad bod and seemingly boring suburban, heterosexual marriage with one child. Later that night, Danny calls Karl and logs into the game. They pick their usual, favorite avatars: Lance, a buff Asian male in martial arts pants and robe for Danny; and Roxette, a toned, sleek Asian woman with blond hair in a short red leather martial arts jacket with undergarments showing for Karl. Karl-as-Roxette handily wins the first fight and then crawls on top of Danny-as-Lance, and they kiss. Freaked out, the two friends exit the game. Over the next few months, after their initial awkwardness, Karl and Danny as Roxette and Lance play the game and repeatedly engage in sexual intercourse. Danny and Karl, embodied as their avatars, have conversations about their sexual relationship and how it feels to them. They affirm the sensuality and intimacy they experience in the game, but wonder if this changes their sexual orientation or gender identity.

Danny and his wife, Theo, are trying to have a child, and Danny realizes his online relationship with Karl-as-Roxette is interfering with his marriage. Danny abruptly ends the relationship, and Karl spirals into a depression, unable to find the same sexual pleasure and intimacy that he has with Danny-as-Lance in the game. Theo, Danny's wife, recognizes that something is wrong but thinks Karl and Danny got in a fight, so she invites Karl over for Danny's birthday a year later. After a very tense dinner, Danny logs into the game that night to confront Karl. They end up having sex again. Danny tells Karl to log off and immediately meet him offline. They try to kiss offline and realize there is no sexual attraction between their offline embodied selves. When

Karl suggests sexuality is different in the game and that they should keep up their gaming sexual relationship, they end up in a fist fight and get arrested. When Danny's wife, Theo, bails him out of jail, he finally tells her about the game and Karl-as-Roxette. The episode ends with a deal. On Danny's birthday he gets to log into Striking Vipers with Karl while Theo goes out without her wedding ring—giving each a twenty-four-hour pass to explore their sexualities.

The questions about sexual orientation and gender identity are not directly answered in this episode. Characters do not affirm their homosexuality or bisexuality as they did in "San Junipero." Instead, the VR experience is used to raise questions about the fluidity of sexual orientation and gender as the users extend themselves into new relationships and embodied experiences. Remember, this game allows them to feel and sense in their offline body all the things happening in the game. The relationship and desire experienced in the game also affects their offline relationships. "Striking Vipers" makes the case that VR as an embodied and relational experience impacts one's sexuality, affecting intimacy, sensuality, behaviors, gender, and orientation. But it does so in ways that do not always conform to the categories of offline experiences where we often perceive orientation and gender to be fixed categories.

What these two episodes suggest is that being part of the story with direct sensory experience, through virtual or simulated reality, may be a way to kinetically, visually, and emotionally change or expand experiences of gender and sexuality. Recent research has shown that virtual reality can change users' ideas and beliefs about the fluidity of gender identity.[27] Researchers brought together two forms of technology—first-person, computer-generated virtual reality, and synchronous physical stimulation—to create a full-body ownership illusion. They posit that "gender identity and the perception of one's own body are tightly connected." The synchronous simulation of experiencing a different gender encoded what researchers call "episodic memories" of gender incoherence in the participants, like Karl, a Black male, experiencing himself as an Asian female in Striking Vipers, including feeling the difference between her programmed experience of orgasm versus his offline experience. In this study, researchers did not simulate sexual behaviors like those in Striking Vipers. Instead, they focused on gender incoherent experiences—having a first-person experience as a male when offline you are female. For the participants, these experiences of being seen and treated as a different gender were not interrupted by other cognitive or emotional responses; instead the participants' gender identities became more fluid. Additionally, researchers reported that "the body-sex-change illusion reduced gender-stereotypical beliefs about [the user's] own personality . . . so that a change in one aspect (gender identification), due to the body-sex-change illusion, affects the other aspects (stereotypical self-beliefs)."[28] This data does not come from a fictional drama like *Black Mirror*, but it does reinforce the experiences narrated in those dramas.

A key feature of these experiences is the multisensory immersion that simulates sensuality and intimacy as key aspects of sexuality. Yorkie, Kelly, Karl, and Danny—through futuristic neural technology—feel, hear, see, smell, and presumably can taste everything happening in the game. Additional research, specifically on sexual arousal in VR, demonstrates the need for multisensory stimuli that take into consideration the room, sounds, haptics or touch, visuals, and smells. This research also contributes to movements against stereotypical representations of body types, unlike "Striking Vipers" and the twentysomething beauty in "San Junipero," by removing visuals of specific people and instead replacing it with more abstract renderings of human bodies. A rich sensory experience that combines offline and online features suggests that "sexual arousal can, therefore, be shared between multiple individuals in VR through the networked experience enhanced with wearable haptic and olfactory devices for all participants."[29] Since sexuality is a holistic experience, only using visuals is not as effective as multisensory experiences.

In fact, there is a new genre of autonomous sensory meridian response (ASMR) erotica, some audiovisual and others audio only, that mimic intimate encounters. Some are specific to sexual experiences, others related to pleasures of eating food or being in nature.[30] ASMR was used in the previously mentioned study to elicit arousal in addition to haptic touch devices and abstract visuals. Researchers and users argue that ASMR, which involves various sensory experiences, helps mitigate against an overreliance on visual simulation in most VR sexual experiences. Yet touch, or the sensation of it through haptic technologies like teledildonics, is also an important feature in sexual arousal, experience, and behaviors in VR settings.[31]

As more research is released on digital sexualities, it becomes increasingly clear that virtual reality technologies allow for anonymity in some cases, deepening of offline relationships in others, gender and orientation fluidity, decreased disease transmission, and potentially safer sexual experimentation with fetishes than offline settings allow.[32] At one point in "Striking Vipers," Karl says to Danny that he tried having sex with most other characters, from human to panda avatars, but none compared to the experience he had with Danny-as-Lance. This raises questions about how offline intimacy, like a best friend relationship, can create a different baseline for intimacy and pleasure in virtual reality. And in "San Junipero," the intimacy developed online grows into relational commitments offline. As we consider the ways humans incarnationally extend themselves through virtual reality and other multisensory devices, we explore ethical opportunities like increasing forms of gender and sexual expression as well as potential harms.

Ethics, Embodiment, and Virtual Reality

Virtual reality allows users to know and be known in sexual ways by connecting or interfacing with other human users and artificial intelligence. These sexual experiences are embodied as they are sensory engagements with the user's body through the extension of their body provided by various forms of software and hardware. That means that where one or two are gathered as Christians in Jesus's name, even in virtual reality, they constitute the communal body of Christ. With this acknowledgment comes Christian ethical commitments, such as the love commandment. The "other" in VR sexual encounters is almost always understood in these contexts as another human. In cases where the other users in virtual reality have known humans behind them, perhaps we would argue for the same ethical treatment of the "other" that is common offline—consent, respect, and mutuality, for example. In such cases we might also consider online sexual experiences cheating if the users are married or coupled offline, as in the case of Karl, Danny, and Theo in "Striking Vipers." But, when the behavior or relationship is human-to-computer, where the avatar is nonreciprocal, a programmed AI, would we argue for different ethical treatment? Should a human be able to do whatever they want to an artificial intelligence avatar or does this "other" require certain ethical treatment as well? These are critical questions we will explore here, and we will continue to engage them in the next chapter when we discuss sex robots.

Virtual reality might offer a variety of sexually healthy and holistic opportunities for sexual behaviors, intimacy, and education. As discussed regarding ASMR research, one can experience sexual arousal and intimacy without physically being with or touched by another person through the use of haptic, olfactory, and VR devices and software. This allows people to explore their sexuality in a context free of social judgment or peer rejection. Thus, VR may also be an educational space to develop communi-

cation skills and become more aware of one's own sexual identity and embodied sexual response preferences. In fact, there are
companies at the intersection of gaming and pornography that
offer VR encounters for sexuality communication and behavioral education.[33] As noted above, VR may also provide a space
to explore sexual fantasies in a safer manner or allow for experimentation with gender and sexuality as we extend ourselves into
virtual spaces.

Yet we also need to be aware of how many VR games prompt
stereotypical visual representations of race, gender, age, and
ability. As addressed in chapter 3, these sexual experiences are
not free of sexual harassment and violence. Human users' experiences of sexual violence in online settings can produce traumatic responses similar to those of offline experiences. Also to
be considered is the moral formation of the actors involved even
if the recipient of the violence is an AI avatar. That's something
we will discuss briefly here—ethical requirements to "others,"
even nonhumans. This is incredibly important to Christian ethics
because sexual expression and behaviors in VR create connections with the self and others in ways that inform one's sexuality,
including aspects of intimacy and sensuality, even if there is no
human on the other end.

The first and most pressing issue to address when considering
Christian sexual ethics may be the lingering stigma that sexual
exploration is morally wrong. This comes from a long history of
considering sexual desire as sinful and from a strict view of sexuality as limited to procreative behaviors in marriage. Throughout
this book, I have argued for a more holistic understanding of sexuality, which includes affirmation of sexual and gender diversity
as part of the human experience, and the active work of iden-
tifying and acknowledging one's desires toward self and others
(erotic attunement). Sexual expression and behaviors in VR allow
users to engage aspects of their holistic sexuality, explore sexual
desires related to orientation, and potentially experience gen-

der in ways that debunk stereotypes. But this is mostly human centered; what about the conversation about incarnation? Isn't it creepy to think of Christ present where two or three are gathered, especially when sexual behaviors are involved? It's creepy only if we have a negative Christian view of sexuality and see God or Jesus as the gatekeeper for sexual purity. That is a teaching that is fairly prevalent in many Christian contexts. Instead, I invite you to consider the ways God revels in the beauty of creation, joins in human joy at worship, or intimately knows every aspect of our being, even the hairs on our head. God wants us to live into our full selves and do so in pleasurable and joyous ways—even, or especially, in ways related to our sexuality.

As we consider virtual reality and extension of bodies, we can look to Jesus both in his instruction in the love commandment, but also in his own practice of extending his presence through others. Jesus's presence is both embodied in his own flesh while incarnated on earth and present in the people with whom we interact on a daily basis. Jesus reminds us that when we heal the sick, feed the hungry, or give drink to the thirsty, we are doing so for him (Matt. 25:35–45). It is also why Jesus is also present when two or three are gathered in his name. In the embodied other, our neighbor, we should recognize Christ or, as we have discussed in other chapters, the *imago dei* (the likeness of God in each of us). That requires living out the love commandment toward the extended self and other, even in virtual reality. As Theodore Jennings, a Christian theologian, writes about the love commandment, "In Christian ethical reflection, therefore we are concerned not with a set of prohibition or legal formulas, but with what conduces to the welfare of those with whom we are in relationship."[34] Related to sexual ethics, he suggests this means we must be attentive to vulnerability and power imbalances, whether they be physical, economic, racial, or gender dependent. Thus, we need a more improvisational approach to ethics as we assess issues of justice and work toward the balance of the commandment to love self, neighbor, and God.

Since we are talking about self and neighbor in relation to the love commandment, we might ask: Does it matter if there is an actual human user behind the avatar in VR sexual encounters? Perhaps to answer this we could lean into the ways in which God is present in all creation, even digital ones. Or, as Scott A. Midson suggests, we could pursue "an alternative approach to neighbourliness that is not fixated on the identity or nature of the other as grounds for assessing whether or not they can be regarded as a neighbour."[35] Midson here is using the test case of moral engagement with a chatbot, not a humanoid avatar. One might argue that technology figured in human form elicits a more ethical response. However, this assumes humans are generally more ethical to other humans, which is not always true. Some individuals might choose to engage in sexual behaviors in VR so they can sexually harm an avatar, something that, done offline, could lead to significant relational and possibly legal ramifications. Midson notes that chatbots can't be affected by hate speech or friendliness, nor can an AI avatar, unless they are programmed to be. That is to say, humans could program an AI to respond to politeness and respect and not respond to disrespect or rudeness. That would at least change the interaction with the technology and elicit certain actions from users. But do we need to do this?

Midson argues that what we really need is to acknowledge that "humans are linked through their conduct both to others and to God."[36] The love commandment requires as much attention to one's own moral formation as to that of a sexual partner and to the community context in which sexuality can flourish or be hindered. Therefore, we should focus on "*interactions*" rather than the substance or function of the other—the AI. This relates directly to the earlier conversation about interfaces when we discussed humans extending themselves with digital technology. Midson suggests we locate the image of God not in human essence or actions, "but rather in our interactions."[37] The interaction is the space of *presencing* Christ, of making the body of Christ present where two or three are gathered. In this sense, sexual ethics follows the love commandment even when the "neighbor" is not a human. Ethics is about relationship and communication, not based on a subject's abilities, such as reason or the substance of a fleshly body.[38]

The ethics of sex in virtual reality depends heavily on the relational network created through human-software-hardware interfaces. Ethical sexual behaviors and expressions in virtual reality evidence and are shaped by the love commandment in and through human-to-human and human-to-computer interactions. We might, for example, suggest that Karl and Danny's relationship on the Striking Vipers game is unethical when it is hidden from Theo, Danny's wife, because the interaction is not living fully into the love commandment by considering all neighbors affected by the relationship. Yet we might say the exact same behavior and relationship is ethical after Karl and Danny tell Theo about their behaviors and they collectively decide on the boundaries. Virtual reality sexual experiences may allow users to reconsider their understanding of sexuality and gender and experience different forms of embodiment limited in offline experiences. These ethical goods are not enough if the context of the user's overall relationships and sexuality online and offline are negatively affected or if others are harmed. The ethical assessment of VR sexuality is not limited to an online context when users are body-centric in offline spaces. A holistic view of sexuality considers offline and online embodiment as intertwined. Thus digital sexual ethics must take into consideration embodied and relational experiences in these unique spaces and their impact on each other.

Everyday Sexual and Digital Ethics for Virtual Reality

- Theology of incarnation—Jesus becoming human—lifts up a body-affirming theology that extends to individuals and to the Christian community as the body of Christ.

  What body-affirming Scriptures or Christian practices do you value? This chapter lifts up creation, incarnation, and Jesus's example of serving others. In what ways do you treat people or computers if you imagine the interaction as evidence of the body of Christ? Respect, consent, inclusion of diverse sexual and gender identities, and neighborliness are a few named in this chapter.

- Understandings of gender and sexuality shift over time with changing bodies as we age and live in different cultural circumstances.

  Remember that online avatars often repeat stereotypes, especially those related to gender and race. Also, we do not yet have robust language of sexual orientations related to digital attraction. Digisexuality helps us think about how we might think of orientation and gender as a spectrum rather than as categories tied to binaries of sex and gender. What types of gender expression or sexual orientation would you use in your VR avatar or find attractive in another?

- Sexual arousal comes in all different forms. Sensuality is a key, and often underdeveloped, aspect of sexuality.

  What kinds of behaviors or sexual desires increase your sexual arousal? How do you engage different senses like smell, taste, and touch when we live in a visual culture? Use VR or other digital technologies like AMSR to help you be more aware of underdeveloped aspects of your sexuality such as sensuality.

- A relationship is a relationship, online or offline.

  If you engage in video games or close relationships online, how might you imagine VR behaviors or relationships impacting your offline relationships and vice versa?

According to Our Likeness

TO DESIGN AND COMPANION WITH A SEX ROBOT

DEVELOPMENT AND USE OF ARTIFICIAL INTELLIGENCE (AI) are growing at a rapid pace, from digital assistants like Siri, Alexa, and Google Assistant to home robotics like vacuums, pool cleaners, and lawn mowers.[1] Both may have AI, but only those that move or assess their surroundings in other ways are considered robots. This burgeoning AI and robotic development affects sex tech from teledildonics to sex robots. Some sex tech incorporates AI to learn from the environment, and some takes different robotic forms.[2] For the purposes of our discussion in this chapter, sex robots are synthetic or nonorganic substitutes for or complements to human-to-human sexual interactions. Sex robots are not a replacement for humans, though some people—such as those who identify as digisexual, as discussed in chapter 4—may prefer a sex robot to a human partner. Because these robots take on humanoid forms, people often interact with them as humans, similar to how people interact with robot dogs as though they were dogs.

For those who desire it, sex robots may offer an opportunity for companionship and physical intimacy. This companionship is not mutually reciprocal, at least not yet. Perhaps machine learning will lead to a sentient AI at some point as the technology keeps developing and there are more and more interested users.[3] Someone might choose to companion with a sex robot for a variety of reasons. Much of the ethical debate has focused on those with intellectual and physical disabilities. Neurodiversity and physical disabilities can lead to discrimination and loneliness, so some ethicists argue that people should be able to find sexual satisfaction and relationship with AI robots. AI robots are already used as caregivers though they often do not have a full humanoid facade.[4] Others argue that using robots and AI to respond to social oppression does not attend to the underlying social causes. It begs the questions of whether Christian ethics, regardless of ability, should be more concerned about sexual behaviors with a nonhuman or the affective need of adults to find closeness and intimacy with safe, nonjudgmental companions.

We need to ask: Is there an ethical difference between having sex with a robot that one considers to be a robot versus having sex with a robot one considers to be a person or at least passing as a person? That is to say, is it the nature of the machine, or the nature of the relationship and the human's understanding of that relationship, that matters most to the ethical deliberations? We interact with robots differently when they are in human form (teledildonic versus sex robot). One might call a sex robot an oversized sex toy, and another might call it a girlfriend. The relationship between user and robot matters, as does the purpose for its creation. As we further consider the theological ideas from chapter 4 about living the love commandment in response to AI and robots, we also need to consider the role of humans as co-creators with God. Genesis 1:26 reminds us that we are made according to God's likeness, and one primary thing that separates us from other animals is our ability to create—to nurture and alter creation. What does the development of sex robots tell us about humanity's co-creational ability, limits, and desires?

On the topic of creating and companioning with sex robots, most people immediately think of science fiction. The movie *Her* follows a man who falls in love with an operating system like Alexa with no physical or virtual body needed. Other films, like *Lars and the Real Girl*, depict relationships with sex dolls that have physical bodies and elaborate back stories but no artificial intelligence. *A.I. Artificial Intelligence*, a rare example of companions that are not female, and the highly sexualized *Ex Machina* feature what this chapter defines as sex robots, because they have a humanoid form, robotic capabilities that mimic human movement or behavior, and artificial intelligence.[5] Perhaps the most popular, recent media portrayal of sex robots is *Westworld*, which we will discuss later in this chapter. Unlike science fiction, most sex robots have only limited capacities for humanlike movement and artificial intelligence, so we will also explore how humans create connections with humanoid forms like sex dolls or other nonhuman AI robotic developments. The present moment begs a reevaluation of the design and programming of sex robots. We will consider how sex robots and the desire for relationship with them may be rooted in a co-creational impulse embedded in human programming and whether or not this is a *loving* relationship.

Sex Robots (Kind Of)

Sex robots are nowhere near the development stages imagined in science fiction fantasies like *Ex Machina* or *Westworld*. Sex robots currently have very limited capacities. The majority of available sex robots include a robotic head affixed to a silicone sex doll body that does not move on its own. Some of the doll bodies include sensors built into or under certain areas of the skin that will trigger sounds, like moaning from the doll's audio capabilities. The only movement comes from the facial expressions or shifting of the head. Using artificial intelligence systems, some sex robots can have basic conversations based on a pre-programmed personality. They may even be able to remember certain facts about the user, like name, favorite meal, color, and so on. Imagine a version of Alexa synched to robotic mechanisms so the lips of the doll move as if the doll were speaking the words. Some companies, such as RealDoll, are very close to this seeming humanlike; others build voice boxes in the back of the robot's head so it speaks in a disembodied manner.[6] So, no, this is not the sex robot of science fiction.[7]

Current development of sex robots shares a few common features. They are "designed with gendered and highly sexual characteristics."[8] The vast majority of them have a female sex, like Harmony RealDoll X; though RealDoll is also working on a male version and has a transgender version of a sex doll that is not a robot. Most technologists would argue that the AI of the robots is genderless until it becomes engaged and is able to learn.[9] However, the voice given to an AI means humans will interact with it in gendered ways and the developers of the algorithms often bring their own racial and gender biases into programming.[10] Some of the preprogrammed personality type options include "funny, shy, charming, cute and submissive."[11] Then the self-learning algorithms that make up the AI engage and learn from interactions with the user, and this can further the bias. But what if that's what a user wants?

Critics of sex robots consider the overly sexualized, gendered, and racialized depictions of females as reasons to ban them. They argue that sex robots are similar to sex workers and even to sex trafficking. The Campaign Against Sex Robots suggests that sex dolls, and specifically sex robots, contribute to negative stereotypes and violence against women and girls.[12] However, there is no conclusive evidence for these claims. And, while sex trafficking is always morally wrong as a form of slavery, there are legitimate ethical arguments from various points of view that both support and criticize sex work, which covers a wide swath of labor—from webcam shows to professional pornography to traditional paid in-person sexual behaviors and many other things in between. There is no evidence that banning sex robots, sex workers, or even pornography will end sexual violence against women and children, which is one of the claims of those trying to ban them.[13] Does the existence of sex robots or pornography foster an appreciable increase in sexual violence? That is not known. Yet overly stereotypical gendered and racialized depictions of bodies seeps into the cultural fabric and so for that reason, later in the chapter, I question the design implications of sex robots and consider ways to make them less stereotypical.

This does raise the question of whether or not someone might make a child-sized sex robot. There is near universal agreement against use of child sex robots.[14] The US Congress banned such manufacturing in 2017.[15] There are some researchers who believe use of computer-generated child images or even child sex robots have potential use in treatment of sex offenders, but this would be in very restricted and controlled settings.[16] Technologists, sexologists, and ethicists generally agree that child sex robots should be banned.

A general narrative of the overly sexualized and eventually violent fembot plays into the fantasies that shape public opinions about sex robots. The highly acclaimed HBO show *Westworld* includes child and adult robots of all shapes, sizes, and racial backgrounds.[17] Yet the narrative still contributes to a gendered, sexualized theme of conquest and exploitation by humans and then eventually by the robots. The two main female characters, Delores Abernathy and Maeve Millay, fit gender and racial stereotypes of an optimistic, sweet, white rancher's daughter and a Black, cynical brothel-owning madam, respectively. The context for the show is a Wild West amusement park that human guests visit. The park offers family activities for those visiting the countryside, or guests can indulge violent Wild West fantasies including rape and murder. The "hosts" or robots are often in the repair headquarters after meeting violent ends. The robots are programmed to forget everything and start each new day from preprogrammed scripts. The host robots are incredibly responsive to guests, and the preprogrammed scripts might take a variety of directions (like Choose Your Own Adventure books that seem to offer choices but really provide multiple paths that eventually lead readers to set conclusions). The robots do not "learn" in the sense of building on past interactive experiences or environments. Or so the viewer and guests of the park are told.

As season one unfolds, which is all we need for the purpose of this discussion, the viewer learns that over time the development of the robots has become more and more humanlike. It includes the programming of reveries—small gestures that are associated with past memories that eventually lead to the robot becoming fully sentient. The lead writer of the park's scripts, Lee Sizemore, offers an impassioned critique of the updates and suggests that park management should stop trying to make the robots more lifelike. He argues that guests come to the park specifically because they know the hosts are robots and that they would not do the same activities or pay the same prices to be with humans. His ideas run up against a corrupt management trying to steal the robot technology and one of the original creators of the robots who wants to give them sentience. The main theme revolves around humanity's ability to be like God and create in its own image. Over the course of season one, the viewer learns that *Westworld* is not so much a commentary on humanoid robots as it is a critique of humanity. The quote from Shakespeare's *Romeo and Juliet*, "These violent delights have violent ends," is shared numerous times during season one. It is reiterated as the hosts take over the park, bridging seasons one and two. I'll end the details here to avoid even more spoilers.

While the artistry of the series is beautiful, the main ethical issues revolve around a number of questions: How humanlike is too much when creating robots? Are humans more likely to violate robots than other humans? And will some humans come to care for robots, maybe even fall in love with them? The sex robot one could buy today differs vastly from those depicted in

Westworld. However, similar stereotypical sexualized bodies and a desired theme of unquestioning loyalty shape the narrative lives of sex robots. Like many science fiction narratives, the robots turn on the humans (pun intended) in ways that exploit pleasure, violence, and greed. Might there be a different, less sensational narrative or, in this case, reality about humans and robots?

Companioning with a Sex Robot

Humans bond with all sorts of robots. A smartphone isn't a robot, but I'm guessing that if we polled most people in the pews it would be the AI device to which they are emotionally closest—yelling at it when it doesn't automatically work, joyous when it pushes a photo memory to notifications, grateful when it gives a calendar reminder just in time, and empty when we reach for a pocket and it it's not there. There are many other AI developments and robots to which people companion, and this might be helpful in society. If that's true, why do we often idealize robots in human form and why do any of them need to be built with aspects of sexuality? Maybe they don't. But when they are, user response or treatment provides important information about humans' co-creational impulses.

Humans form attachments to objects, especially to those in human and animal form. The character of those attachments and their influence on human-to-human relationships are key to understanding the ethics of sex robots.[18] Children are often given baby dolls to nurture. This practice allows children to engage in a variety of relational expressions. Sometimes the child befriends a doll; other times they parent it. They can insist that siblings and parents treat the doll as real. They express the doll's emotions and make it talk. The child is learning about relational responsibility, expression of feelings, and adaptation in a social context. Society also generally believes that kids should grow out of the use of stuffed animals or dolls. Yet plenty of grown-ups still have stuffed animals, keep security items, or have collectibles (adult approved toys). These objects bring pleasure and comfort; people form attachments to them that cannot be limited to age.[19]

Humans also form attachments to other nonhuman animals. We talk about the personalities of pets and carry on conversations with them. Pets are nonhuman animals, but humans are constantly interacting with them through human standards of behavior. We anthropomorphize them. All one needs to do is read the comment section of a famous dog or cat Instagram account to see the various ways we relate to pets as though they were human. We do this because we relate to everything out of our own frame of reference—being human! The attachments we form with nonhuman animals raise questions about mutuality and control. These relationships are unequal, and yet we often say we love our pets.

One of the most successful care robots brings together the human experience of loving stuffed animals and pets. PARO is a therapeutic robot.[20] It is a fuzzy baby seal that responds to touch with movement and seal noises. PARO is used mostly in senior facilities and was first studied in patients with Alzheimer's disease. PARO's presence led to increased communication by and with patients, and to reduced anxiety. It's not clear if the patients think PARO is an animal or a robot, but that's really not the point. They adjusted positively to PARO's presence and formed an attachment that led to an increase in better human relations with their caregivers and others in their small groups.

Similarly, the robot Pepper helps humans engage their emotions.[21] Pepper is not a cute, cuddly animal bot. Pepper is a human-sized bot that looks like a traditional robot, with a smooth white enameled exterior, a head with large screen eyes and tympanic circles for ears, jointed arms and hands, a large screen in the center of its chest, and a roller box bottom to navigate rooms. Pepper was initially designed to move about health care facilities to check in on patients; its machine learning AI enables it to have conversations with patients primarily related to emotional issues. This is an incredible feat given that affective recognition is difficult for humans and thus programming emotionally intelligent AI is difficult. Pepper is now used in other environments like museums and schools. It is specifically designed as a robot with features that humans can relate to, but not to be mistaken as being human. This provides the opportunity to connect without the expectation of exact human-to-human communication standards.

James McBride wonders how Christian communities will integrate robots into faith communities. He argues, as most technologists do, that robots will become ubiquitous in our lives in only a few decades. McBride focuses on care and companion robots, which could include sex robots, with whom we might have the greatest connection. While his questions are mostly about robot converts and congregation members, he also discusses the question of whether certain value systems or religious beliefs could or should be programmed into robots according to an owner's preference. McBride believes that as robots become more like humans, exhibiting caring behaviors for example, we will treat them like social agents and expect other moral functions from them.[22]

The examples of care robots demonstrate intentional design choices that allow for human connection with robots and create distance from human forms. However, there are good reasons to design robots in human form. Kate Devlin, author of *Turned On: Science, Sex, and Robots* (which is about a lot more than sex robots) notes, "We have built environments suitable for humans: human-sized doors and paths, stairs the right size for human legs and shelves within the reach of our arms. It would be quite handy to have robots that could fit into our environment."[23] Yet, just as Sizemore asks in *Westworld*, we have to question what happens when the humanoid form begins to deceive us and we forget they are robots? For example, why make a robot with a sexuality?

Isn't that only a human quality? No. All animals have a sexuality, and humans already read sexuality and gender into places it doesn't exist. We would do this with robots even if it wasn't explicitly programmed.

A robot probably will never experience all five aspects of sexuality discussed throughout this book—gender and orientation, intimacy, sensuality, sexual and reproductive health, and sexual behaviors. Let's suppose a robot could be programmed for sensuality or intimacy. Would its experiences of those two things be the same as human's or would they be designed responses to human cues such that a user would interpret the actions as intimacy or sensuality? For example, Spanish engineer Sergi Santos designed a sex robot named Samantha with sensors inside a sex doll body and an AI system in her head. She can learn responsively as the sensors read feedback from penetration causing orgasm, a form of programmed sensuality. AI learning aspects of Samantha are preprogrammed with what I would identify as certain ethical characteristics. She can learn patience based on her attention algorithm or will shut down (go into dummy mode) if experiencing overly aggressive touch. This mimics emotional responsiveness and closeness or intimacy. These interactions suggest a reciprocity required in her use that is not always present in other sex robot designs.[24] Samantha's reactions are not the same as consent, a freely made choice in human terms. She is programmed to follow certain rules, and thus the responses are chosen from a preset range of options. But that doesn't make them unethical. We have to understand the limitations of the robot and determine ethics for a robot from their own design, not from human capacities.

Humans can and do bond with robots, even when they are in forms we fully recognize as robotic and not human. The humanoid form of sex robots also means they have certain aspects or identifiers of sexuality, including being able to engage in sexual behaviors, being given a sexual orientation and gender identity through design and use, expressing sensuality through sensor technology, and expressing intimacy through certain AI features and modes. These features of sexuality help humans companion with sex robots. Especially expressions of pleasure, sensuality, and intimacy can create the feeling of relational interactions.[25] Yet a sex robot's sexuality is not the robust and integrative sexuality experienced by a human. This is an important recognition as companionship with a sex robot is very clearly human-to-robot, not human-to-human, interaction.

Technological Concerns and Social Limitations of Sex Robots

A sex robot is a robot. Right now, the models are not that sophisticated. What if mass produced sex robots eventually have the various technologies like those embedded in Samantha's design, not full sentience as in *Westworld*? The sex robot could hold a conversation with some level of emotional analysis; shut down when abused; and learn certain interactional features like patience, sexual excitement, and responses to touch across its body. These would be incredible advances in sex robot technology and perhaps are not that far off. Even in their current state, very far from full sentience, proponents of sex robots believe they provide companionship, sexual pleasure, may supplement other human relationships, and could serve educational purposes. In terms of limits, a sex robot cannot reciprocate with the same mutuality as a human. Some believe this lack of reciprocity will lead humans to use sex robots in place of human relationships because they are "easier" and can be controlled. This leads to questions about robot rights and the limits that should be placed on creators and users to avoid exploitation and abuse like what happens in *Westworld*. It also signals a concern about the security of a user's data as they interact with networked machine learning AI in the form of a sex robot.

A primary concern about sex robots relates to digital surveillance and safety.[26] A sex robot, like all of our other digital devices, needs to feed data back to a larger system in order to progressively learn on the scale and at the speed to which users are accustomed. For example, mobile phones are equipped with speech to text functions that allow a user to voice text. This technology has gotten so good we might even be frustrated when it occasionally misspells a name or uses *there* instead of *their*. It is not only a single user's interaction with their device that hones this technology; it's billions of users feeding data back to Google or Apple. The same is needed from a network of sex robots to improve their conversational skills. Would the data only be language and conversations, or would sex robots eventually share locational GPS data or visual spatial data from a home? User data security is a major digital technology issue, and it is increasingly concerning given the number of networked machine learning devices in homes. Thus this issue is relevant to all devices, not only sex robots, though the nature of the information shared via sex robot data collection may (or may not) be more sensitive than what's recorded by a digital assistant.

Author of *Love and Sex with Robots* David Levy is well known for his predictions that robots will replace human sexual partners in only a few decades.[27] He argues that robot relationships are easier and give humans a sense of control when human-to-human romantic and sexual relationships are often difficult. Levy's approach is sometimes characterized by critics as gender-biased and heterosexist. We can hear the male stereotype that a female sex robot is easier than a bossy, emotional woman. But it is not only heterosexual men who are interested in robotic companions. Using an Inmoov pattern, a woman in France 3D printed a robot companion that she wants to marry.[28] This robot, however, is clearly a robot more like Pepper, described above, than a RealDoll like Harmony. In these cases, and those of many sex doll owners, the person has a preference for a robot as a sexual partner. But this doesn't mean they are against all relationships with humans.

Some readers might be familiar with the work of Sherry Turkle on the social impacts of digital technology. She argues there is an inverse relationship between humans' growing reliance on digital technology and their eschewing of human connection.[29] Directly applying this hypothesis to sex robots is a stretch in comparison to something like social media usage. Kate Devlin does not fully agree that most people in the future will choose to marry a robot, but she also disagrees with the hypothesis that companioning with a robot means the user will become antisocial and stop interacting with humans. One could argue that having a positive and pleasurable relationship with a sex robot might help someone be more socially confident. Devlin turns to the iDollator community, sex doll owners, to explore this point. Davecat, a doll owner and star of a documentary on sex dolls, reports being more social because of the community he has met through owning sex dolls.[30] He notes all the other human relationships he has at work, with family, and with other friends. He doesn't feel like he is missing something or that his romantic and sexual relationship with dolls is less than a human relationship. Additionally, the iDollator community, as a prototype for the way users may treat sex robots, suggest that relationships are marked by respect, care, pleasure, and, yes, love.[31]

Davecat is clear when he speaks about his relationships with sex dolls that he knows they are not human, regardless of the back stories and personalities he has created for them.[32] He wants to be in relationship with them because they are dolls, not humans. As concerning as it might sound for a person to want a sexual relationship with a doll and not a human, would it not be more concerning if he considered the doll to be the same as a human? The recognition of the distinction matters. In this sense, he and many other sex doll and robot owners avoid the theological conundrum that Noreen Herzfeld describes. As a theologian, she considers the danger of robot relationships as trying to make an I-Thou out of an I-It—meaning, "The danger lies in our using relationship with a robot as a template for relationship with other persons."[33] This is because a robot is a robot, not a human. As already noted above, a sex robot with some eventual upgrades may be able to reciprocate with an emotional reaction or express sexual pleasure when stimulated. Yet that is not the same as unprogrammed, spontaneous, altruistic responses in humans.

The fact that robots are different than humans may mean that some readers feel it is never ethical to have a sexual relationship with a robot or they will simply see it as a life-size sex toy but not a relationship. For those who do see a robot as an "other" with whom one could potentially form a relationship, we need to address the question of robot rights and the limits that should be placed on creators and users to avoid exploitation and abuse. While we have addressed some of the benefits and harms to the human user and to society, what about the robots? Some critics argue that sexbots would create a new form of slavery. However, the argument rests on using human standards as the starting point; they assume robot desires and agency are corrupted by human use and control. Yet society does not regularly worry about large AI manufacturing machines being enslaved. They are built and programmed for a purpose. Like sex robots, they "come into existence wanting what they do."[34] Still, to avoid abuse that can happen to any machine, we might change the model of sex robots from owned to commissioned, as philosopher Steve Petersen suggests. The sex robot might be designed specifically for an individual user and thus the social and legal responsibility of its human companion. If the fictional narrative of *Westworld* has taught viewers anything, it's that neglecting and abusing robots perpetuates violence as a key developmental feature in robots, humans, and society at large. We need a social system that recognizes robot rights on the basis of their purpose and impact on society.

William F. Schulz and Sushma Raman make a compelling case for how we might extend rights beyond humans toward animals, artificial intelligence (AI), and nature.[35] They employ a constructivist approach that avoids binding rights to a "natural" order with humans at the top of the hierarchy. Rather, they ask how the treatment of animals, the developmental direction of AI, and human co-existence with nature contribute to a good society. The establishment of a right in the constructivist approach is also not reliant on a "thing" possessing some particular quality in order to be granted a right. For example, animals, robots, and earth do not possess the same levels of consciousness, ability to feel pain, displays of altruism, and so on as humans do. Instead, Raman and Schulz ask the reader to consider judging how a thing best fulfills its purpose as a criterion for applying rights. Here is where the authors start to draw limits so not all animals or AI or things in nature would have the same rights. There would be criteria related to flourishing and interdependence with humans that distinguish between "rights claiming" animals, AI, and nature.[36]

The first part of a constructivist litmus test for a right is not that difficult to affirm in most cases: "On balance, if this right is recognized, will it contribute to a greater degree of human dignity? Will it make for a good society in which I want to live?" Sex robot rights enhance certain human characteristics we want to promote, such as treating others—machine or natural—with respect and care. That in turn adds to the good of society. However, the second portion of the test requires constraining the powerful in favor of those with less: "Or, alternatively, how will the denial of this right harm a segment of our population and hence damage our claim to be a good society?"[37] Denying rights to sex robots could lead to humans using them in violent ways that could impact the moral formation of individuals and negatively shift cultural understandings of sexuality.[38] We need social action and legal regulation toward robot rights. These rights are interdependent with human rights and a collective view of the good society, including protecting user data.

Creating Our Companions in Sex Robots

Humans are relational beings. "A God who makes relational covenants and in whose image we are made suggests that we are 'hard-wired' to do the same, to seek everywhere a partner with whom to relate, not just among ourselves, human with human, but also with someone or something different from us."³⁹ As I asked at the start of this chapter: Is it the nature of the machine or the nature of the relationship and the human's understanding of that relationship that matters most for ethical deliberations? I am convinced that it is the relationship that matters the most. Indeed, it begs the question of why Christians are more concerned about sexual behaviors with nonhumans than about the affective need of adults to find closeness, intimacy, and pleasure with safe, nonjudgmental companions. Such relationships may mirror back to humanity greater clarity about sexuality and a varied, diverse understanding of love and companionship similar to how our createdness mirrors back something of the differences and desires of God, our Creator. We may ultimately decide that sex robots do not need to be in humanoid form or at least not in overly stereotypical gendered forms.

The development of sex robots is not only for sexual behaviors, at least not those robots that take a humanoid form. The expressed need or desire for intimacy and sensuality as aspects of sexuality are part of the desire for a humanoid form with AI. For some users, sex robots will supplement their human-to-human sexual relationships; for others they will provide their primary sexual relationships. There is no evidence, however, that sex robots will lead users to end human relationships. Sex robots may serve educational and companionship purposes in addition to providing sexual pleasure. For example, engagement in sexual behaviors with a touch-sensitive and responsive sex robot might help humans become better at or more comfortable with sexual behaviors. Some individuals who have affective communication difficulties could either choose to companion with a robot to reduce loneliness and offer an opportunity for sexual pleasure or they may learn from interactions with an emotionally intelligent AI robot and eventually be more comfortable communicating with other people. In a study of users who were asked to interact with Harmony, an AI RealDoll X, most users talked positively about the impact of a relationship with her, including reduced loneliness, having a nonjudgmental presence, and experiencing physical touch or closeness.[40]

Digital technologies of all varieties are often used to mitigate disabilities. Often these technologies are seen as neutral tools that people use to solve a socially defined deficit. Jana Bennett, writing on disability and theology, suggests we see both technology and disability from a social perspective rather than machine- or tool-based approach. Disabilities are determined by a set social context; thus adding an assemblage like a wheelchair gives one the opportunity to navigate a world created for walking humans. Dominant standards for social interactions and the requirement of a committed relationship in order to experience companioned sexual behaviors excludes a lot of people. While the requirement of a committed relationship is shifting, the social interactive expectations are strong. Bennett instead suggests we focus on the relationship (between human, technology, and disability), and in particular she wonders how our relationships with technologies incarnate Christ and yield the neighborliness of the love commandment.[41] These questions and theological themes are similar to those raised at the end of chapter 4.

When we consider whether a human could love a robot, we are likely to immediately assume it has to be a romantic love equivalent to what we imagine as the ideal of a marriage. Yet most marriages never reach that ideal of a love that is mutual, reciprocating the desire to altruistically promote one another's well-being and flourishing. Some philosophers argue that in a human-to-human loving relationship there is a potential for active, mutual reciprocity toward well-being and flourishing, even if we never achieve it or cannot sustain it.[42] They suggest this is not even possible with robots, given that they don't have a well-being to promote and that flourishing for them would simply mean living out their program.[43]

I am struck by Kate Devlin's question, "Who are we to judge that love has to be reciprocated to be valid?"[44] The obsession with this type of love is mostly due to a separation of "types of love" in the Christian tradition related to philia (friendly), eros (desiring or romantic), and agape (neighborly). I'm not convinced this separation is possible in everyday life, and if we were to argue for one as primary it would be agape. As noted in chapter 4, Scott Midson argues that "humans are linked through their conduct both to others and to God."[45] The love commandment requires as much attention to one's own moral formation as to that of a sexual partner and to the community context in which sexuality flourishes or is hindered. Therefore, we should focus on the relationship, not the nature of the parties involved. Midson suggests we locate the image of God not in human essence or actions, "but rather in our interactions,"[46] which become the presence of Christ. In this sense, we are to follow the love commandment even when the "neighbor" is not a human.

When it comes to relating to robots as our neighbors, Anne Foerst suggests the human desire to create robots and have relationships with them is a continuation of the creational drive God bestowed in humans. She reflects on the medieval Jewish tradition in Kabbalah of creating *golems*, figures with magical powers to act and behave like humans, linking this to the modern creation of humanoid robots. She notes that *golem* is the same "word used in Psalm 139 for the thing we are shaped into in our mother's womb."[47] She suggests that through creation of this kind of object, as an act of worship and made animate through the power of a prayer held in its mouth, humans participate in God's creation. For Foerst, because we are created in the image of God, "we have the drive to create, to repeat God's acts of creation. The very desire of God to create humans as partners is inside us."[48] Foerst argues that as we build robots in our image, we will build them with the best of humanity's traits. "So the very act of robot building is an act of loving: a loving recognition of who we are and a loving rendition of who we might want to become."[49] Most of the robots in *Westworld* do not yield the best of humanity's traits, but those traits do seem to be part of Sergi Santos's purpose in creating Samantha.

As co-creators, perhaps more creativity is in order. A number of feminist sexologists and technologists suggest that someone with a clitoris may find a nonhumanoid robot more pleasurable. Sexual stimulation of a clitoris requires varied movements beyond simple penetration. Perhaps the reader noticed that I wrote that sentence avoiding the use of *woman* or *man* and focusing on stimulation of a certain body part. This might be the kind of thinking that's needed if the purpose of a sex robot is primarily sexual pleasure. Most currently designed sex robots and dolls recreate gendered and racialized stereotypes of female and some male forms. If we keep making robots in humanoid forms, will we move away from racial, ethnic, gender, age, and body-shape biases? Probably not!

Instead, creators might design forms that maximize sexual pleasure and physical comfort. Kate Devlin and her graduate students often hold hackathons to dismantle sex toys and create better ones in an effort to spark the imagination of the next generation of creators.[50] Others have also noted that just as sex-positive feminism has intervened in the pornography industry, it could also help with robot design and programming.[51] This includes working to "ensure better content [i.e. depictions or representations of female (and male) sexuality in robotic form], better processes (i.e. more female voices included in the production and distribution of sexbots), and better contexts (i.e. social environments and conversations surrounding the consumption and use of sexbots)."[52] I would go even further and suggest that we don't need to work in gender binaries at all. Some individuals will want a cisgender robot, but others will not. God's creation is a spectrum of sexual and gender diversities, and if we are to be creators inspired by God's work we should seek such diversity as well. Additionally, design needs include safeguards related to data and surveillance as well as to physical and emotional treatment of the robot's AI and components.

Robots are not yet what we imagine in science fiction, and perhaps that is a good thing. It gives us time to rethink those narratives in light of the best qualities of humanity rather than those that stereotype and reduce the full diversity of God's creation. Maybe this is the point at which we need to stop and explore different forms. Sexuality and gender expressions are changing and expanding rapidly, unfolding the fullness of God's creation. Recreating overly gendered female bodies may respond to a very small slice of sexual desire and companionship. Not only is that a bad business model; it's a diminishment of co-creational possibilities and may lead to fairly stunted relations with our new neighbors.

Everyday Sexual and Digital Ethics with Sex Robots

- Imagine your ideal sexual partner.

  Each of us has a unique sexuality created by God even if some of the main features of one person's sexuality overlap with that of others. We are not all attracted to the same type of person or want to engage in the same sexual behaviors. What do you desire in a sexual partner—what looks, behaviors, orientations, identities? What do these desires tell you about your own understandings or biases regarding sexuality or gender?

- Consider key features of a healthy sexual relationship.

  Relationships are difficult, but they also teach us a lot about ourselves. Relationships can require vulnerability, communication, sacrifice, growth, and so much more. What things that make human-to-human sexual relationships difficult or messy would you like to give up? What would you miss if your primary relationship was human-to-robot? What humanlike ethical characteristics should be built into robot design and programming? What makes one a good neighbor in a relationship?

- Many people struggle to find intimacy and companionship either temporarily or for their whole lives.

  How can human-to-robot interactions supplement or help humans who have differing needs, from health care settings to private homes? Can sexual pleasure be an ethical part of this companionship?

- "How do I love thee? Let me I count the ways." We show love in many different, diverse ways to humans, pets, and some inanimate objects.

  What are the various kinds of love that describe these relationships for you? How would you describe the type of love needed for a relationship to include sexual behaviors?

Youth Study Guide

THE YOUTH STUDY GUIDE IS DESIGNED to engage teens and young adults in the subject matter of this book. Each section provides a brief summary of specific chapter-related theological, sexuality, and digital technology information to serve as a primer for the adult leader. However, the guide assumes the leader has read each chapter, equipping them with a thorough overview of the topic. The study guide can be customized to suit the group's age, knowledge level, and meeting times. There are additional suggestions for further study as well.

I provide a variety of resources that may help a youth leader prepare for using this discussion guide. Many youth leaders have never received holistic faith-based sexuality education. There are free online resources through the Yale Youth Ministry Institute to help equip youth leaders who want to teach about sexuality. While there is no section to accompany the introduction, I recommend a leader consider an introductory conversation or educational experience that helps participants define sexuality and gain a better understanding of sexual ethics.

The following resources will help ground youth leaders in a holistic faith-based sexuality education approach and provide resources to conduct a sexuality education program assessment. There are also curriculum materials to help youth and young adults define their own sexual ethics along the lines found in the introduction of this book.

- Quest for Life, Sexuality Education Videos and Curriculum: https://yaleyouthministryinstitute.org/?s=sexuality
- Ministry with LGBTQ+ Youth: https://yaleyouthministryinstitute.org/resources/hot-button-topics/ministry-with-lgbtq-youth/

CHAPTER 1

In the Image of God: Youth Study on Digital Pornography

Instructor Summary

Sexuality Education Information

Pornography can refer to a variety of sexually explicit materials, with an emphasis on visual media including short videos, photos, live web-camming, full-length movies, and so on. Pornography is also a legal and artistic category. Some sexuality educators distinguish between pornography, erotica, and sexually explicit educational materials; the latter two do not include violence or exploitation. Most youth do not make these distinctions, but it is helpful to raise the issue that there is a variety of sexually explicit material online and not all of it is violent or exploitative. Most youth initially explore pornography out of curiosity and to seek information that is lacking due to inadequate sexuality education. They continue to explore sexually explicit material for all kinds of reasons: ongoing education, boredom, entertainment, pleasure, exploration, fantasy, control, consumption, an emotional outlet, and connection. Youth also create self-produced pornography through sexting, which is a widespread sexual behavior, though the explicitness of the photos or messages differs based on a young person's comfort and the length of the relationship.

Experts differ on whether there is such a thing as addiction to pornography. A small group of mostly heterosexual, male users do exhibit compulsive behaviors and experience negative outcomes like sexual dysfunction, poor body image, and poor relationship communication. However, these may indicate strong correlations rather than causation. For example, one might have poor relationship skills, turn to pornography for sexual gratification, and then uncritically adopt pornography scripts as the way to talk to a sexual interest on a date. If sexually explicit materials are not empowering and enhancing their sense of sexuality and sexual relationships, users should stop or find different, more affirming sources.

Digital Technology Primer

As noted, sexually explicit online materials can be produced by amateurs or professionals. Digital technologies provide the means of production, publication, and distribution on a widespread basis that was not formerly available. These materials are also more widely accessible, low-cost or free, and able to be privately consumed via a personal digital device. The opportunity for direct consumption and production means many more youth of today view and make pornography than did those of past generations. Digital technology also leaves a data trail, so privacy is never absolute and often one's behavior can be easily identified via search history.

Theological Engagement

Throughout Scripture, we are reminded that we are created by God and made in God's image. This positive affirmation means that the whole of every person, including their body and their sexuality, is indeed good, or very good as we hear in Genesis. Youth can find it difficult to affirm their whole selves, including their bodies, at a time when so much is changing due to puberty and when peer pressure is at its highest levels. Adults in their lives need to constantly repeat, "You are a child of God. God loves you no matter what." This doesn't mean we can engage in any actions we choose, but it is an affirmation that counters peers and media telling them they are not good enough.

God created our sexuality and our bodies to experience sexual attraction and pleasure. That is part of the goodness of our creation. We affirm that goodness when we appreciate our own or another person's beauty or communicate our sexual attraction in a way that acknowledges the image of God in another. We need to cultivate our Christian belief that our bodies are to be valued and respected as part of God's good creation and extend that to those we encounter via image or video online. The image of God in each of us is affirmed when we are seen as whole, complex human beings; and when we are loved for who we are, not for who someone else wants us to be. This means that, before sharing in intimate connections either offline or online, we should learn to stop and evaluate our desires and decisions.

Youth Gathering Outline

Scripture Text

In order to remind youth that they are made in God's image, begin with a prayer or meditation like the one below, based on Genesis 1:27 and 31. Invite the youth to get comfortable, maybe put their feet on the ground, relax their arms, or close their eyes and breathe in and out. Then, slowly and intentionally read the following meditation written to be gender inclusive. It can be read two or three times, while reminding the youth to breathe in and out with each phrase.

[breathe in] God created humankind in God's image.
[breathe out] And it was very good.
[breathe in] In the image of God, we are created.
[breathe out] And it was very good.
[breathe in] God saw everything God made.
[breathe out] And it was very good.

Sexuality, Digital, and Theological Perspectives
on Digital Pornography

Provide a summary of the information above and from the chapter either through talking points or a short question and answer time for the youth. Remember to focus on the difference between consumption and education or appreciation in the use of sexually explicit materials.

Questions for Ethical Expansion

- How do you identify the difference between using sexually explicit material for education versus consuming it like popular media?
- Consider people's comments to a posted video or photo. How do you know when someone is honoring another person's body rather than objectifying it?
- In what ways does learning more about your body and sexuality deepen your relationship with self, a partner, and God? In what ways could it separate you from self, others, and God?

Exercise

Invite the group to participate in an exercise to apply what they have learned. They can either do this individually, in pairs, or in small groups.

Create an app that displays a few questions and a positive theological message that would pop up on a person's digital device before they log into any site with sexually explicit material, e.g., "Before you go any further, [insert message here]."

Or, consider a twist on this exercise and have the youth create an app that displays a question and positive theological message before they send a sext.

Remember, the point is not to block the behavior of the user, but to use a popup or login screen to make them stop and evaluate their actions so the user can make their own informed decisions and practice reflecting on their values.

For more ideas see: "Defining Pornography" Youth Study at The ThoughtfulChristian.com: https://www.thethoughtfulchristian.com /Products/TC5208/defining-pornography.aspx

---

CHAPTER 2
I Sought One Whom My Soul Loves: Youth Study on Online Dating

Instructor Summary

Sexuality Education Information

Relationships are difficult to navigate (period). Youth and young adults are constantly renegotiating their relationships with family, friends, coaches, teachers, employers, and romantic or sexual interests. Learning to balance all these relational responsibilities while also figuring out who and what one wants to be in the world can be overwhelming. Sexual desire and attraction are aspects of sexuality. As our sexuality unfolds, we need to be attuned to our desires and feelings so we don't act without reflection and intention. In other words, a sexually healthy person knows the difference between feelings and actions.

Many youth default to particular scripts about what hooking up or dating is supposed to look like. Unfortunately, these scripts do not lead to flourishing or healthy relationships; they can be based on gender stereotypes and unrealistic expectations of happiness. Changing the relationship script to a values-based relationship requires practice and constant negotiation. It provides a higher standard for healthy relationships that are built on communication and shared expectations.

Digital Technology Primer

Youth and young adults primarily communicate through digital modes—texts, snaps, DMs, video calls, and so on. The designs of these systems shape new forms of communication, for instance using imagery in place of words for communication. Photos, emojis, and short videos appear more than text. Text and language have also changed either through creating new words or shortcuts like acronyms for faster communication. Thus expectations change regarding communication formats. When one texts or DM's someone, they expect a fairly immediate response, whereas a snapchat only requires a daily response to keep up a streak of communication. And even though video calling like FaceTime is more frequently used than phone calls, there is no expectation that both parties will sit and look at each other the whole time. Parents and other leaders should invite youth to discuss the new and changing expectations of communication in order to develop critical engagement and discernment.

We are available 24/7 because of our digital devices, and without some predefined boundaries, youth and young adults may not have the space they need to disconnect and reflect. Digital communication may also substitute for in-person time (caused by a number of factors), but every relationship needs both. The geographic and time collapse of digital communication also means youth can create relationships with people they may never meet in person. That expands their potential social circles but also opens them up to possible harms. There is also the likelihood that many of these young adults will use dating or meetup sites, so it is important to discuss the harms (addressed in the next study) and discuss the impacts of visual and biographical representations of self and others as a way to select a partner.

Theological Engagement

Youth and young adults need a solid foundation of Christian values upon which to judge the quality of all relationships, especially those that are romantic or sexual. I often turn to 1 Corinthians 13:4–9 as a reminder of which values lead to love and which do not. Building a relationship based on values like honesty, kindness, patience, trust, commitment, and equality requires constant work and communication.

A values-based starting point helps mitigate the hazards of a visually driven culture. And yet we are sexual beings who find other people attractive, often because of their appearance or how they comport themselves in a group. It's wonderful to appreciate the attractiveness of another person. God created our sexuality, and sexual desire and attraction are part of sexuality for most people. While there are some major categories for sexual orientation, like heterosexual or homosexual, each person has a unique sexuality that extends to those whom they find attractive. Spending time considering what makes someone attractive creates space to look for qualities that push against social standards of beauty or dismiss aspects of personality or shared interests as possibly more important when considering a partner.

Youth Gathering Outline

Scripture Text

Most churches avoid reading Song of Songs, considering it too explicit or erotic for church contexts. However, it is part of Scripture and should be valued like other portions. Otherwise we send a message that faith has nothing to do with sexual desire or attraction. Provide a short background for the youth on the Song of Songs in the Bible and read a passage. Consider an inclusive reading of

Love's Dream, Song of Songs 3:1–5. Invite feedback on what it's like to search for someone you love, to desire them and make that known publicly, and to wait until the time is right.

> **Love's Dream**
> Upon my bed at night
>   I sought the one whom my soul loves;
> I sought them, but found them not;
>   I called them, but they gave no answer.
> "I will rise now and go about the city,
>   in the streets and in the squares;
> I will seek them whom my soul loves."
>   I sought them, but found them not.
> The sentinels found me,
>   as they went about in the city.
> "Have you seen the one whom my soul loves?"
>   Scarcely had I passed them,
>   when I found the one whom my soul loves.
> I held them, and would not let them go
>   until I brought the one into my mother's house,
>   and into the chamber of her that conceived me.
> I adjure you, O children of Jerusalem,
>   by the gazelles or the wild does:
> do not stir up or awaken love
>   until it is ready!

Sexuality, Digital, and Theological Perspectives on Online Dating

Provide a summary of the information above and from the chapter either through talking points or a short question and answer time for the youth. Remember to focus on shared values in a relationship and the need to attune ourselves to sexual desire and attraction.

Questions for Ethical Expansion

- Was this the first time you have read from Song of Songs? If so, what is your reaction—surprise, concern, questioning, discomfort?
- What standards of beauty form your default attractions or preferences? Do these standards of appearance keep you from meeting certain people? What role does social media play in creating or reinforcing these standards?
- Do your online profiles accurately represent who you are? Would someone looking at them know you are Christian? What values—honesty, respect, care, support—undergird the way you communicate through DMs, snaps, and other replies?
- Describe a healthy, loving relationship. Who do you know that has such a relationship? What do you think are key factors that make it healthy and loving?

Exercise

Relationship health and strength come from shared values and good communication. Using a board or poster paper, create two columns and have the youth brainstorm positive and negative behaviors and values in a relationship. Reading a passage like 1 Corinthians 13:1–9 can help to get the list started.

Then ask each youth to pick their top five values that they believe create a healthy relationship.

Next, have youth pair up and give an example of each value in action in a relationship related to digital media. Run this discussion like speed dating. Each pair talks for five minutes, taking turns giving an example; then they find a new person to talk to for their next example.

Debrief with a large group feedback session on how it felt to talk about these values and connect them to examples of everyday digital communication.

This exercise helps youth and young adults practice various types of communication as well as identify their core values.

For more ideas, see "The Dating Game: How to Survive the Roller Coaster" and "Forming a Healthy Relationship" at TheThoughtful Christian.com: https://www.thethoughtfulchristian.com/Products /TC5058/the-dating-game.aspx and https://www.thethoughtful christian.com/Products/TC5060/forming-a-healthy-relationship .aspx.

_____

CHAPTER 3

Love Does Not Delight in Evil: Youth Study on Online Sexual and Dating Violence

Instructor Summary

Sexuality Education Information

Teens and young adults experience intimate and dating violence in their own relationships and witness such violence in the adult relationships around them. Sexualized and intimate violence happens across all racial, economic, age, educational, and geographic demographics. While heterosexual men, gay men, and lesbians are less likely to be victims of dating violence, it still happens, just at a lower rate than for heterosexual women and trans people. Patterns of violence and harassment include manipulation, control, and coercion, all of which are forms of power over another person. Sexual violence and harassment can happen once or be ongoing. It can cause lasting sexual trauma for victim-survivors that can affect future relationships as well as other aspects of their lives like school, extracurricular activities, work, and family relationships.

We all need to be educated to identify the signs of sexual abuse and harassment and forms of emotional, spiritual, and physical violence in relationships. Recognizing intimate violence and responding to it are very different things. Many victim-survivors find it very difficult to end abusive relationships because of the manipulating behavior of the perpetrators and the fear of escalating the violence. Having friends and family who are supportive is extremely important to strengthen the foundation of any relationship, but it is often crucial to escaping an abusive one.

Digital Technology Primer

Youth and young adults experience various forms of sexual violence and harassment online. The nature of some online violence and harassment means it is difficult to stop and expose. The communication can be public or within a friend group, but it is more likely direct, personal, and private. Knowing how to document online abuse and harassment is a key aspect of digital literacy, as is being able to report and block individuals. Often perpetrators are someone the victim-survivor knows, so that can make it more difficult to disentangle from them in online spaces because of multiple networks of connection. In addition, digital software and hardware can be used by a perpetrator to track or surveil a partner's whereabouts, search history, or messages. Some educational groups online, like LoveIsRespect, mentioned at the start of chapter 3, have resources to help with digital literacy and dating violence prevention.

Theological Engagement

The general taboo about sexuality in faith communities creates silence, making it difficult to talk about sexual abuse and harassment. This silence, plus theological messages about suffering and self-sacrifice, can confuse youth and young adults into thinking they are responsible for abuse and harassment directed at them. The gender dynamics of self-sacrificing messages can compound problems for young women who are told to be submissive and serve the men in their lives. These compounding factors result in a very harmful mix. Christian teachings need to distinguish between Jesus's voluntary suffering and the involuntary nature of abuse and harassment. In congregations where the whole community takes responsibility for ending violence and abuse this silence is broken, and leaders and teachers are accountable for perpetuating harmful theological teachings. This includes a youth or young adult responsibility to be an upstander to a friend if they suspect abuse or harassment.

Youth Gathering Outline

Scripture Text

Remind the youth that 1 Corinthians 13:1–7 is often read at weddings. However, Paul, the writer of the letter, wasn't speaking only to a romantic couple. He was talking to the church community at Corinth. His guidance on love is for all relationships—communal and individual. If you already read 1 Corinthians 13 in the chapter 2 youth study, this will be a refresher. Read the passage as a meditation or prayer, with each group member reading a phrase.

> Love is patient;
> love is kind;
> love is not envious or boastful or arrogant or rude.
> It does not insist on its own way;
> it is not irritable or resentful;
> it does not rejoice in wrongdoing, but rejoices in the truth.
> It bears all things,
> believes all things,
> hopes all things,
> endures all things.

Sexuality, Digital, and Theological Perspectives on
Online Sexual and Dating Violence

Provide a summary of the information above and from the chapter either through talking points or a short question and answer time for the youth. Remember to focus on gender inclusive theologies that combat the harms of sexual and intimate violence and encourage youth to be upstanders rather than bystanders in online and offline relationships.

Questions for Ethical Expansion

- Paul's letter expresses what love is and is not. We often learn in church that Jesus was the ultimate example of love. He treated everyone with respect, even those who were diseased or social outcasts. Some people believe we should sacrifice and suffer when following Jesus's example. But his death on the cross was voluntary. What's the difference between voluntary and involuntary suffering? Does Jesus's sacrifice mean we should stay in abusive or harassing relationships or does it suggest the opposite?
- A lot of online harassment is rooted in gender discrimination and homophobia. How does your faith help you stand up for equality, inclusion, diversity, and belonging for all God's people?
- What are the digital means to being an upstander—standing up against sexual violence and harassment when you see it? How can you respond, block, or protect yourself from having your accounts attacked or harassed?
- In what ways does your faith community advocate for people in need? What could your faith community do to support domestic and sexual violence programs in your area?

Exercise

Share the teenage vignette from chapter 3 or find a more recent and relevant example from the news or a popular TV show. Ask the students to identify the behaviors that contribute to sexual harassment and talk about how they would respond or support a friend in a similar situation.

After the discussion, reread the following lines from 1 Corinthians—"Love is not envious or boastful or arrogant or rude. It does not insist on its own way; it is not irritable or resentful; it does not rejoice in wrongdoing"—and remind the youth that the definition of sexual harassment and violence includes manipulation, coercion, and control.

Give them safe resources online and in your community related to sexualized violence prevention and reporting before they leave the gathering.

For further resources on youth and sexual and gender violence, see "Responding to Violence in Jesus' Time and Now" at TheThoughtful Christian.com: https://www.thethoughtfulchristian.com/Products /TC5107/responding-to-violence-in-jesus-time-and-now.aspx.

For a popular young adult novel on deepfakes and online harassment, see Sarah Darer Littman, *Deepfake* (New York: Scholastic Press, 2020).

CHAPTER 4

Where Two or Three Are Gathered: Youth Study on Sexuality and VR

Instructor Summary

Sexuality Education Information

Sexuality includes a variety of aspects, even though we often reduce it to behaviors and orientation. Intimacy and sensuality are key to a holistic understanding of sexuality and are experienced in a variety of relationships online and offline. Embodiment requires a rich engagement with the senses and emotional, spiritual, and physical closeness to self, others, and God. Neglecting these aspects of sexuality to solely focus sexual behaviors stifles the fuller development of one's sexuality. Similarly, categories related to gender and sexual orientation can be confining and politically charged. Virtual reality is a space to "walk a mile in another's shoes" to better understand gender discrimination as well as explore one's gender and orientation and gain a fuller understanding of one's sexuality.

Digital Technology Primer

Virtual reality technology extends human perception, sense, and embodiment beyond the boundary of human skin. Virtual reality connects or interfaces users with one another or with avatars, sometimes with artificial intelligence. These experiences are embodied as they are sensory engagements using various forms of software and hardware. Some are completely immersive, using simulation goggles and haptic wearables, and others offer third-person experiences through video games. Youth and young adults probably have more experience with virtual reality or 3D worlds than most adults, given the variety of video games they have played while growing up. Video games or separate online community worlds like Second Life are the primary locations for VR sexual relationships and behaviors.

Theological Engagement

Youth and young adults are often taught about the incarnation as a way to talk about Jesus's divinity. But this misses a major consideration of how believers can connect with Jesus, especially when it comes to sexuality. First, if Jesus is human, he experienced puberty, sexual attraction, vulnerability, sexuality, and intimacy; that's all part of being embodied. Second, Jesus's incarnation—his taking on a human body and showing up in the bodies of our neighbors as we serve them—grounds the theology of the church as the body of

Christ. The body of Christ is not one person. While the body as in-carnational is central to being human, the boundary of our skin is not where incarnation ends. It is the conduit or interface for relationship, especially in the digital realm. We meet self and others through our bodies, and at that point of interaction God is present. This brings material meaning to the love commandment. Love of God, self, and other isn't an abstract concept. In the direct embodied actions of loving oneself and others, God is present. The opposite is also true: In sinning against oneself or another, God is rejected. This can be a difficult theological concept to explain. But when we make it visual, actually seeing the interface or interaction in the meeting, we can visualize God being present even in virtual reality.

Youth Gathering Outline

Scripture Text

Begin with a drawing exercise based in the meditative readings of the following two passages: Matthew 22:30–44 and Matthew 18:20. Have the youth realistically or abstractly draw the relationships expressed in the passages. Consider playing background music. Read the passages a few times, taking a break between readings as needed. Save the discussion of these drawings for the exercise portion of this youth study.

Sexuality, Digital, and Theological Perspectives
on Sexuality and VR

Provide a summary of the information above and from the chapter either through talking points or a short question and answer time for the youth. Remember to review the diverse aspects that comprise sexuality, including intimacy and sensuality, as part of embodied ex-periences online and offline.

Questions for Ethical Expansion

- In what ways do you treat people or computers if you imagine the interaction as evidence of the body of Christ?
- What types of gender expression would you use in your VR avatar or find attractive in another's?
- How could engaging multiple senses help you better understand your sexuality? For example, we live in a highly visual culture; how does this affect the role of taste, touch, smell, or hearing related to sexual expression? Have you used or would you use VR or other digital technologies like AMSR to help you be more aware of under-developed aspects of your sexuality like sensuality?
- Do you have relationships online with people you have never met offline? Or do you feel connected to or invested in a video game character even though you know they are not a person? In these examples, how do these "virtual" relationships shape your views on intimacy and sensuality related to online relationships?

Exercise

For this exercise, return to the opening Scripture exercise; or if you did not do it, begin with it here. Invite a reflective discussion on the drawings, asking questions like: How did you visually depict relationship? What do God and Christ look like in your drawings? Were the default images all human forms or did some youth use abstract symbols, colors, or movements to create connection and image the divine?

Challenge the youth to develop a virtual reality space where God in any trinitarian form is present and moving among the avatars as they interact. Have them draw or narrate one scene from this game that they can share with the group.

For more resources on youth and digital representations see "What Digital Story Are You Writing?" at TheThoughtfulChristian.com:

https://www.thethoughtfulchristian.com/Products/TC5098/what
-digital-story-are-you-writing.aspx.

_____

CHAPTER 5
According to Our Likeness:
Youth Study on AI Robot Companions

Instructor Summary

Sexuality Education Information

Companionship is a human need, though companionship does not
need to match stereotypical relationships like marriage. Many people
cultivate intimacy with a best friend or a family member, or even a
beloved pet or a special place in nature. While many people may
eventually want a romantic or sexual partner, that's not always possible. Some individuals choose sex dolls or robots as their companions.
Sex robots could have an educational purpose. Practicing with a sex
robot may enhance performance of sexual behaviors or be used as a
form of treatment for those who experience sexual dysfunction or
past trauma. The current overly sexualized, gendered, and racialized
humanoid forms of sex robots, however, contribute to negative sexual stereotypes that could perpetuate oppressive social standards.
The design guide for the development of sex robots should account
for the diversity of human sexualities.

Digital Technology Primer

Current design and sales of sex robots are far from the sci-fi depictions seen in movies. There are a number of different types of care

or companion robots that integrate AI technology. Youth and young adults have experience with many devices that include AI, like their smartphones and in-home assistants and robots like vacuums and toys. Designers create particular robotic features to match up with user desires and needs. That's no different when it comes to companion robots. It's important to acknowledge that companion or sex robot designs mimic cultural stereotypes of beauty and femininity because that is what designers think users want.

Theological Engagement

God creates humans "according to God's likeness." This endows humans with the gift of being co-creators and wanting to be in relationship with one's creation. We are created as relational beings to love self, others, and God—similar to the triune nature of the Trinity. That desire and need for relationship shapes how humans interact and what they desire. When humans co-create, as with God, the product of the creation tells us about the creator. As we move into rapid development of AI and robots, we need to consider the rights and responsibilities this work requires. Human technological creations like companion robots reflect back to humanity what we value in others and in relationships.

Youth Gathering Outline

Scripture Text

Being made in God's likeness means humans reflect aspects of God. Open with a prayer that thanks God for creating us, acknowledges the responsibility we have as co-creators and caretakers of creation, and asks for God's wisdom and mentorship in that work.

Remind the youth that Genesis 1:26 says humankind is made in God's image and according to God's likeness. Ask them what parts of humanity remind them of God or reflect aspects of God.

Sexuality, Digital, and Theological Perspectives
on AI Robot Companions

Provide a summary of the information above and from the chapter
either through talking points or a short question and answer time
for the youth. Remember to focus on the co-creational responsibility
that humans have in technological development, especially of AI
robots that mimic people.

Questions for Ethical Expansion

- What are the devices you and your family use that have artificial intelligence or robotics? These digital devices are more prevalent than
  we think. How do you interact with these devices? Are you polite in
  giving commands? Do you expect perfect response or completion
  of tasks? Have these devices replaced past activities? If so, what's
  gained and lost by using them? Connecting AI and robotics to everyday use helps us see the ways we are shaped through interaction
  with them.
- What things make human-to-human sexual or romantic relationships
  difficult or messy that you would like to give up or change? What
  would you miss if your primary relationship was human-to-robot?
- What humanlike ethical characteristics should be built into robot
  design and programming? In other words, what makes a robot a good
  neighbor in a relationship?
- "Love" is a word that carries lots of meanings. How would you describe
  the type of love needed before you engage in sexual behaviors in a
  relationship?

Exercise

Begin with a brainstorming session about robot characteristics depicted in popular media, especially characteristics of humanoid robots
from movies, books, comics, and more. Then share the main features

of various companion robots that are already being used and that are mentioned in chapter 5—such as PARO, Pepper, Samantha, and so on. Keep track of all these features on a board or poster paper.

Individually or in small groups invite the youth to prioritize the kinds of features that would be highest on their list if they were designing a companion robot as co-creators with God. What are the traits of humanity they want represented? Have the youth verbally or visually share their designs.

Notes

Acknowledgments

1. Philip Wogaman, *Christian Ethics: An Historical Introduction* (Louisville: Westminster John Knox, 1993), 222.

Introduction

1. These aspects of sexuality come from a visual definition of sexuality referred to as the "circles of sexuality." The original version designed by Dr. Dennis M. Dailey uses five overlapping categories: sensuality, intimacy, sexual identity, sexual health and reproduction, and sexualization. More recent versions separate sexual identity into sexual orientation and gender identity, add sexual behaviors and practices, and remove the sexualization circle replacing it with an overlapping circle covering the other five labeled power and agency. For further detail on these aspects of sexuality and the history of the circles, see Heather Corinna, *S.E.X., the Second Edition: The All-You-Need-to-Know Progressive Sexuality Guide to Get You through High School and College* (Boston: DeCapo Press, 2016), 17. Corinna also explains the history and context of the circles of sexuality in "Sexuality: WTF Is It Anyway," found online at http://www.scarleteen.com/article/bodies/sexuality_wtf_is_it_anyway.

2. Bonnie J. Miller-McLemore, "Embodied Knowing, Embodied Theology: What Happened to the Body," *Pastoral Psychology* 62 (2013): 750.

3. Margaret Farley, *Just Love: A Framework for Christian Sexual Ethics* (New York: Continuum, 2006), 231, and Miguel De La Torre, *Liberating Sexuality: Justice between the Sheets* (St. Louis, MO: Chalice, 2016), 22–23.

4. Kate Ott, *Sex + Faith: Talking to Your Child from Birth to Adolescence* (Louisville: Westminster John Knox, 2013).

5. Cristina L. H. Traina, "Erotic Attunement," in *Professional Sexual Ethics: A Holistic Ministry Approach*, eds. Patricia Jung and Darryl Stephens (Minneapolis: Fortress, 2013), 44. My own engagement with Traina and her work is also present in Kate Ott, "Orgasmic Failure: A Praxis Ethic for Adolescent Sexuality," in *Theologies of Failure*, ed. Roberto Sirvent and Duncan B. Reyburn (Eugene, OR: Cascade Books, 2019).

6. Cristina L. H. Traina, *Erotic Attunement: Parenthood and the Ethics of Sensuality between Unequals* (Chicago: University of Chicago Press, 2011), 141.

7. Traina, "Erotic Attunement," 47.

8. Traina, *Erotic Attunement*, 203.

9. Traina, *Erotic Attunement*, 217.

10. Traina, *Erotic Attunement*, 243.

11. For more on ethics as art, play, or improvisation, see Marvin M. Ellison, *Making Love Just: Sexual Ethics for Perplexing Times* (Minneapolis: Fortress Press, 2012); Theodore W. Jennings Jr., *An Ethic of Queer Sex: Principles and Improvisations* (Chicago: Exploration Press, 2013); John Wall, *Ethics in Light of Childhood* (Washington, DC: Georgetown University Press, 2010); and Kate Ott, *Christian Ethics for a Digital Society* (New York: Rowan & Littlefield, 2019).

12. Marcella Althaus-Reid, *Indecent Theology: Theological Perversions in Sex, Gender and Politics* (London: Routledge, 2000), 146.

Chapter 1

1. Janis Wolak, Kimberly Mitchell, and David Finkelhor, "Unwanted and Wanted Exposure to Online Pornography in a National Sample of Youth Internet Users," *Pediatrics* 119 (2007): 247–57.

2. See the recent article from the American Psychological Association, "Task Force on the Sexualization of Girls, Report" (2010), http://www.apa.org/pi/women/programs/girls/report-full.pdf; Jochen Peter and Patti M. Valkenburg, "Adolescents' Exposure to Sexually Explicit Online Material and Recreational Attitudes toward Sex," _Journal of Communication_ 56, no. 4 (December 2006): 639–60; Dale Kunke et al., _Sex on TV 4, A Kaiser Family Foundation Report_ (KFF, November 2005), https://www.kff.org/other/sex-on-tv-4-report/.

3. Heather Berg, _Porn Work: Sex, Labor, and Late Capitalism_ (Chapel Hill: University of North Carolina Press), 2021. Additionally, Tristan Taormino, a producer of pornography, covers changes in the pornography industry on a number of her podcast episodes on _Sex Out Loud with Tristan Taormino_, https://sexoutloud.libsyn.com.

4. J. Michael Bostwick and Jeffrey A. Bucci, "Internet Sex Addiction Treated with Naltrexone," _Mayo Clinical Proceedings_ 83, no. 2 (2008): 229.

5. This scenario is a composite created from the research reviewed and conducted in Janet K. L. McKeown, Diana C. Parry, and Tracy Penny Light, "'My IPhone Changed My Life': How Digital Technologies Can Enable Women's Consumption of Online Sexually Explicit Materials," _Sexuality & Culture: An Interdisciplinary Quarterly_ 22, no. 2 (2018): 340, doi:10.1007/s12119-017-9476-0.

6. Bonnie J. Miller-McLemore, "Embodied Knowing, Embodied Theology: What Happened to the Body," _Pastoral Psychology_ 62 (2013): 750.

7. Miller-McLemore, "Embodied Knowing, Embodied Theology," 744.

8. Rob Rhea and Rick Langer, "A Theology of the Body for a Pornographic Age," _Journal of Spiritual Formation & Soul Care_ 8, no. 1 (2015): 91.

9. Karen Peterson-Iyer, "Mobile Porn? Teenage Sexting and Justice for Women," _Journal of the Society of Christian Ethics_ 33, no. 2 (2013): 93–110. On page 104, Peterson-Iyer writes, "While at times well-intentioned, purity advocates perpetuate the same social rubrics that guide girls to understand their own sexual desire as a source of shame and embarrassment [rather] than encouraging girls to understand themselves as moral agents and sex as a moral and deliberate choice."

10. Christine E. Gudorf, *Body, Sex, and Pleasure: Reconstructing Christian Sexual Ethics* (Cleveland: Pilgrim Press, 1995), 51.

11. Cristina L. H. Traina, *Erotic Attunement: Parenthood and the Ethics of Sensuality* (Chicago: University of Chicago Press, 2011). Traina's proposal thus diverges from the historical tradition of procreative ethics originally tied to the writings of St. Augustine of Hippo and still largely promoted by Roman Catholic doctrine today. See her discussion, in chapter 3, of Augustine's ideal of self-mastery that has generated a tradition of Christian sexual ethics that treats others as an object, considers male desire as dangerous—overpowering the will and reason—and sets up women as the objects upon whom sex is enacted. She also points to the fact that the procreative ethic "drains much of the ethics of relationship from the ethics of sexuality" (88). See also p. 241 of her book.

12. Linda Williams, *Hardcore: Power, Pleasure, and the "Frenzy of the Visible"* (Berkeley: University of California Press, 1989), 34–36.

13. See most trafficked websites list in 2020 at https://www.sem rush.com/blog/most-visited-websites/.

14. Mark Regnerus, David Gordon, and Joseph Price, "Documenting Pornography Use in America: A Comparative Analysis of Methodological Approaches," *Journal of Sex Research* 53 (2015): 1–9, doi:10 .1080/00224499.2015.1096886.

15. For a more popular article commentary on this study see the online Psychology Today blog from 2018 on "New Data from the World's most Popular Porn Site" (Pornhub) at https://www.psychologytoday .com/us/blog/all-about-sex/201803/surprising-new-data-the-world -s-most-popular-porn-site, accessed November 2020.

16. Franklin O. Poulsen, Dean M. Busby, and Adam M. Galovan, "Pornography Use: Who Uses It and How It Is Associated with Couple Outcomes," *Journal of Sex Research* 50, no. 1 (2013): 72–83, doi:10.1080 /00224499.2011.648027.

17. Destin N. Stewart and Dawn M. Szymanski, "Young Adult Women's Reports of Their Male Romantic Partner's Pornography Use as a Correlate of Their Self-Esteem, Relationship Quality, and Sex-

ual Satisfaction," *Sex Roles* 67 (2012): 257–71, https://doi.org/10.1007/s11199-012-0164-0.

18. Aleksandra Diana Dwulit and Piotr Rzymski, "Prevalence, Patterns and Self-Perceived Effects of Pornography Consumption in Polish University Students: A Cross-Sectional Study," *International Journal of Environmental Research and Public Health* 16, no. 10 (May 27, 2019): 1861, doi:10.3390/ijerph16101861.

19. McKeown et al., "'My IPhone Changed My Life,'" 340.

20. McKeown et al., "'My IPhone Changed My Life,'" 345.

21. McKeown et al., "'My IPhone Changed My Life,'" 349. See also, Aristea Fotopoulou, "The Paradox of Feminism, Technology and Pornography: Value and Biopolitics in Digital Culture," in *Feminist Activism and Digital Networks: Between Empowerment and Vulnerability* (London: Palgrave Macmillan UK, 2016), doi: 10.1057/978-1-137-50471-5_3.

22. Beáta Bőthe, Marie-Pier Vaillancourt-Morel, Sophie Bergeron, and Zsolt Demetrovics, "Problematic and Non-Problematic Pornography Use among LGBTQ Adolescents: A Systematic Literature Review," *Sex Addiction Current Report* 6 (2019): 478–94, doi:10.1007/s40429-019-00289-5.

23. Penny Harvey, "Let's Talk about Porn: The Perceived Effect of Online Mainstream Pornography on LGBTQ Youth," in *Gender, Sexuality and Race in the Digital Age*, ed. D. Nicole Farris, D'Lane R. Compton, and Andrea P. Herrera (Cham: Springer, 2020), doi:10.1007/978-3-030-29855-5_3.

24. The original documentary was aired in 2013 on British television. See https://www.imdb.com/title/tt3238490/. The documentary has been redone to focus on updated research that connects brain chemistry changes, evolutionary biology, and digital technology. The primary claim is that "rebooting" the brain through a break from online pornography will alleviate sexual dysfunction caused by excessive use of online pornography. See the Your Brain on Porn website, https://www.yourbrainonporn.com/videos/your-brain-on-porn-how-internet-porn-affects-the-brain-2015/.

25. Aleksandra Diana Dwulit and Piotr Rzymski, "Prevalence,

Patterns and Self-Perceived Effects of Pornography Consumption in Polish University Students: A Cross-Sectional Study," 1861.

26. Kirsten Weir, "Is Pornography Addictive?," *Monitor on Psychology* 45, no. 4 (April 2014), http://www.apa.org/monitor/2014/04 /pornography.

27. Joshua B. Grubbs et al., "Moral Incongruence and Compulsive Sexual Behavior: Results from Cross-Sectional Interactions and Parallel Growth Curve Analyses," *Journal of Abnormal Psychology* 129, no. 3 (2020): 266–78, doi:10.1037/abn0000501. See also a response to a study by Jonathan Merritt, commissioned by the Barna Group, on pornography use, "Pornography: A Christian Crisis or Overblown Issue?," in *Religion News Service*, January 20, 2016. https://religion news.com/2016/01/20/christians-pornography-problem/.

28. Meagan J. Brem et al., "Problematic Pornography Use and Physical and Sexual Intimate Partner Violence Perpetration among Men in Batterer Intervention Programs," *Journal of Interpersonal Violence*, November 21, 2018: doi:10.1177/0886260518812806.

29. Daisy J. Mechelmans et al., "Enhanced Attentional Bias towards Sexually Explicit Cues in Individuals with and without Compulsive Sexual Behaviours," *PLoS One* 9, no. 8 (August 25, 2014): doi:10.1371/journal.pone.0105476.

30. Shane W. Kraus et al., "Compulsive Sexual Behaviour Disorder in the ICD-11," *World Psychiatry* 17, no. 1 (2018): 109–10, doi:10.1002/wps.20499.

31. Antonio García-Gómez, "Teen Girls and Sexual Agency: Exploring the Intrapersonal and Intergroup Dimensions of Sexting," *Media, Culture & Society* 39, no. 3 (April 2017): 402, doi:10.1177/0163443716683789.

32. Many current laws in the United States treat sexually explicit materials of a minor, whether self-produced or not, as child pornography, carrying a felony class offense and possible "sex offender" designation. In recent years, states have tried to amend these laws though some still use them to prosecute those who are secondary sexters and send an image without the permission of the original sender. García-Gómez, "Teen Girls and Sexual Agency," 392.

33. Swathi Krishna, "Sexting: The Technological Evolution of the Sexual Revolution," *Psychiatric Times* 36, no. 12 (December 2019): 24–25.

34. Shafia Zaloom, "'Hey, WYD?': Inside Teen Sexting," Girls Leadership, August 2019, https://girlsleadership.org/blog/hey-wyd -inside-teen-sexting/.

35. García-Gómez, "Teen Girls and Sexual Agency," 393–96.

36. See media examples shared by Karen Peterson-Iyer and her discussion of the disproportionate social and emotional risks to heterosexual girls related to sexting in "Mobile Porn? Teenage Sexting and Justice for Women," *Journal of the Society of Christian Ethics* 33, no. 2 (2013): 93–110.

37. Sheri Madigan et al., "Prevalence of Multiple Forms of Sexting Behavior among Youth: A Systematic Review and Meta-analysis," *JAMA Pediatrics* 172, no. 4 (2018): 327–35, doi:10.1001/jamapediatrics .2017.5314.

38. Peterson-Iyer, "Mobile Porn?," 97–98.

39. García-Gómez, "Teen Girls and Sexual Agency," 391–407.

40. Panayiota Tsatsou, "Gender and Sexuality in the Internet Era," in *The Handbook of Gender, Sex, and Media,* ed. Karen Ross (Oxford: Wiley & Sons, 2012), 517.

41. Tsatsou, "Gender and Sexuality in the Internet Era," 531.

42. Sylvain C. Boies, Gail Knudson, and Julian Young, "The Internet, Sex, and Youths: Implications for Sexual Development," *Sexual Addiction & Compulsivity* 11 (2004): 343–63, doi:10.1080/10720160490902630.

43. Jaco J. Hamman, "The Organ of Tactility: Fantasy, Image, and Male Masturbation," *Pastoral Psychology* 67 (2018): 632, doi:10.1007 /s11089-017-0797-6.

44. Ana Bridges and Patricia Morokoff, "Sexual Media Use and Relational Satisfaction in Heterosexual Couples," *Personal Relationships* 18 (2011): 562–85, doi:10.1111/j.1475-6811.2010.01328.x.

45. Miguel De La Torre, *Liberating Sexuality: Justice between the Sheets* (St. Louis: Chalice Press, 2016), 18.

Chapter 2

1. *The X-Files,* season 3, episode 6, "2Shy," directed by David Nutter, written by Chris Carter and Jeff Vlaming, first aired November 3,

1995, Ten Thirteen Productions and 20th Century Fox Television. *The Circle*, season 1, released January 1, 2020, Studio Lambert, https://www.netflix.com/title/81044551.

2. Michael J. Rosenfeld, Reuben J. Thomas, and Sonia Hausen, "Disintermediating Your Friends: How Online Dating in the United States Displaces Other Ways of Meeting," *Proceedings of the National Academy of Sciences* 116, no. 36 (2019): 17753–58, doi: 10.1073/pnas .1908630116.

3. Alexander Weinstein, "Comfort Porn," in *Universal Love: Stories* (New York: Holt, 2020), 49–77.

4. Maria Stoicescu, "The Globalized Online Dating Culture: Reframing the Dating Process through Online Dating," *Journal of Comparative Research in Anthropology and Sociology* 10, no. 1 (2019): 21–32.

5. "The Virtues and Downsides of Online Dating," Pew Research Center, Washington, DC (February 6, 2020), https://www.pewre search.org/internet/2020/02/06/the-virtues-and-downsides-of -online-dating/.

6. "Here's What Will Get You Kicked Off of Bumble," Bumble, accessed April 21, 2021, https://bumble.com/en-us/the-buzz/what-will -get-you-kicked-off-bumble.

7. Begonya Enguix and Elisenda Ardevol, "Enacting Bodies: Online Dating and New Media Practices," in *The Handbook of Gender, Sex, and Media*, ed. Karen Ross (West Sussex: Wiley & Sons, 2012), 503.

8. For a further technological explanation of the variety of influences that shape participation in online dating applications and considerations of the data cultures including platform design, algorithms, marketing, and social bias, see Kath Albury et al., "Data Cultures of Mobile Dating and Hook Up Apps: Emerging Issues for Critical Social Science Research," *Big Data & Society* 4, no. 2 (December 2017): doi:10.1177/2053951717720950.

9. Enguix and Ardevol, "Enacting Bodies," 502.

10. "So, What Is the Science behind eHarmony?," eHarmony Blog, accessed April 20, 2021, https://www.eharmony.com/blog/science -behind-eharmony/.

11. For a personal example of this phenomenon read Aymann Ismail, "'I'm Not a Bigot Because I Prefer a Certain Kind of Person,'" _Slate_, August 03, 2019, https://slate.com/human-interest/2019/08/asian-men-racism-dating-bias-man-up.html.

12. Teresa Milbrodt, "Dating Websites and Disability Identity: Presentations of the Disabled Self in Online Dating Profiles," _Western Folklore_ 78, no. 1 (Winter 2019): 67, https://search-proquest-com.yale.idm.oclc.org/scholarly-journals/dating-websites-disability-identity-presentations/docview/2161594741/se-2?accountid=15172.

13. Dan Slater, _A Million First Dates: Solving the Puzzle of Online Dating_ (New York: Penguin, 2014).

14. Kate Julian, "The Sex Recession," _The Atlantic_, December 2018, 78–94.

15. For more data on this point see Maria Stoicescu, "The Globalized Online Dating Culture," 21–32, and Katie Ann Schubert, "Internet Dating and 'Doing Gender': An Analysis of Women's Experiences Dating Online," PhD diss. (University of Florida, 2014), https://search-proquest-com.yale.idm.oclc.org/dissertations-theses/internet-dating-doing-gender-analysis-womens/docview/1645427732/se-2?accountid=15172.

16. Julian, "The Sex Recession," 78–94.

17. Enguix and Ardevol, "Enacting Bodies," 512.

18. Body shaming and inhibition are two of the minor negative sexuality-related outcomes reported by ex-evangelicals and others who were part of purity culture movements in their teens. See Linda Kay Klein, _Pure: Inside the Evangelical Movement That Shamed a Generation of Young Women and How I Broke Free_ (New York: Atria Paperback, 2018).

19. Julian, "The Sex Recession," 86.

20. See Todd A. Salzman and Michael G. Lawler, "Cohabitation and the Process of Marrying," in _The Sexual Person: Toward a Renewed Catholic Anthropology_ (Washington, DC: Georgetown University Press, 2008), 192–213.

21. See Rosemary Radford Ruether, *Christianity and the Making of the Modern Family* (Boston: Beacon, 2001).

22. Justin R. Garcia, Chris Reiber, Sean G. Massey, and Ann M. Merriwether, "Sexual Hookup Culture: A Review," *Review of General Psychology* 16, no. 2 (2012): 161-76.

23. Ellen Lamont, Teresa Roach, and Sope Kahn, "Navigating Campus Hookup Culture: LGBTQ Students and College Hookups," *Sociological Forum* 33 (2018): 1000-1022, doi:10.1111/socf.12458.

24. Marvin Ellison, *Making Love Just: Sexual Ethics for Perplexing Times* (Minneapolis: Fortress, 2012), 76.

25. See Renita Weems, *What Matters Most: Ten Lessons in Living Passionately from the Song of Solomon* (New York: Walk Worthy Press, 2004).

26. David M. Carr, *The Erotic Word: Sexuality, Spirituality, and the Bible* (New York: Oxford University Press, 2003), 117.

27. Carr, *The Erotic Word*, 119.

28. Carr, *The Erotic Word*, 140.

29. Audre Lorde, "Uses of the Erotic: The Erotic as Power," in *Sister Outsider: Essays and Speeches* (Freedom: The Crossing Press, 1984), 56.

30. Carr, *The Erotic Word*, 149.

31. For more on the function of communities of accountability and support, especially in the predominantly Black church contexts, see Monique Moultrie, *Passionate and Pious: Religious Media and Black Women's Sexuality* (Durham: Duke University Press, 2017). Moultrie's analysis shows how sexuality ministries are complex, negotiated spaces that yield diverse sexual ethics regardless of the dominant theological approach.

Chapter 3

1. See Love Is Respect at http://www.loveisrespect.org.

2. For example, see M. Shawn Copeland, *Enfleshing Freedom: Body, Race, and Human Being* (Minneapolis: Fortress, 2009); Marcella Althaus-Reid, *From Feminist Theology to Indecent Theology: Readings on Poverty,*

Sexual Identity and God (London: SCM, 2008); and James H. Cone, _The Cross and the Lynching Tree_ (Maryknoll, NY: Orbis Books, 2014).

3. Traci C. West, "Ending Gender Violence: An Antiracist Intersectional Agenda for the Churches," _Review & Expositor_ 117, no. 2 (May 2020): 199–203, https://doi.org/10.1177/0034637320924015, 200. Also see West's other writings for a more in-depth discussion of the intersections of Christian theology, sexual abuse, gender, and racism. See Traci C. West, _Solidarity and Defiant Spirituality: Africana Lessons on Religion, Racism, and Ending Gender Violence_ (New York: New York University Press, 2019); Traci C. West, _Disruptive Christian Ethics: When Racism and Women's Lives Matter_ (Louisville: Westminster John Knox, 2006); and Traci C. West, _Wounds of the Spirit: Black Women, Violence and Resistance Ethics_ (New York: New York University Press, 1999).

4. I discuss this intersection related to youth group and Sunday School teen sexuality education in "What Do Good Friday, Teens, and Dating Have in Common?," _Social Justice Leadership Project Blog_, May 2019, https://www.drew.edu/theological-school/2019/05/01/what-do-good-friday-teens-and-dating-have-in-common/.

5. Marie M. Fortune, _Sexual Violence: The Sin Revisited_ (Cleveland: Pilgrim Press, 2005).

6. Delores S. Williams, _Sisters in the Wilderness: The Challenge of Womanist God-Talk_ (Maryknoll, NY: Orbis Books, 1993).

7. Rosemary Radford Ruether, "Feminist Metanoia and Soul-Making," in _Women's Spirituality: Women's Lives_, ed. Judith Ochshorn and Ellen Cole (Philadelphia: Haworth Press, 1995), 38.

8. Ruether, "Feminist Metanoia and Soul-Making," 33.

9. See chapter 3 in Kate Ott, _Christian Ethics for a Digital Society_ (New York: Rowman & Littlefield, 2019).

10. Ruether, "Feminist Metanoia and Soul-Making," 34.

11. Ken Stone, "What the Homosexuality Debates Really Say about the Bible," in _Out of the Shadows into the Light: Christianity and Homosexuality_, ed. Miguel A. De La Torre (Danvers, MA: Chalice Press, 2009).

12. Marvin M. Ellison, _Making Love Just: Sexual Ethics for Perplexing Times_ (Minneapolis: Fortress, 2012), 16.

13. Jia Tolentino, "The Rage of the Incels: Incels Aren't Really Looking for Sex. They're Looking for Absolute Male Supremacy" in *The New Yorker*, May 15, 2018, https://www.newyorker.com/culture /cultural-comment/the-rage-of-the-incels.

14. Tolentino, "The Rage of the Incels."

15. Rob Rhea and Rick Langer, "A Theology of the Body for a Pornographic Age," *Journal of Spiritual Formation & Soul Care* 8, no. 1 (2015): 99.

16. Thomas Renkert, "Subaltern, Supraaltern, and the Digitally Grievable Life," *Cursor_ Zeitschrift für explorative Theologie*, April 20, 2021. https://cursor.pubpub.org/pub/renkert-subaltern-supraaltern.

17. Hans Block and Moritz Riesewieck, *The Cleaners, A Documentary*, Gebrueder Beetz Filmproduktion, Grifa Filmes, Westdeutscher Rundfunk (WDR), 2018. https://www.bpb.de/mediathek/273199/the -cleaners.

18. Caroline Kitchener, "When Cyberstalking Crosses Borders," *The Atlantic*, December 5, 2017. https://www.theatlantic.com/member ship/archive/2017/12/when-cyberstalking-crosses-borders/547523/.

19. Michelle Cottle, "The Adultery Arms Race," *The Atlantic*, November 2014, https://www.theatlantic.com/magazine/archive/2014 /11/the-adultery-arms-race/380794/.

20. Hanna Rosin, "Why Kids Sext," *The Atlantic*, November 2014, https://www.theatlantic.com/magazine/archive/2014/11/why-kids -sext/380798/.

21. Emily A. Vogels, "The State of Online Harassment," Pew Research Center, January 13, 2021. https://www.pewresearch.org/in ternet/2021/01/13/the-state-of-online-harassment/.

22. See recommendations and policy responses to issues of privacy and surveillance at the Electronic Privacy Information Center, "Domestic Violence and Privacy," https://epic.org/privacy/dv/. To find the host of a website see http://whois.icann.org.

23. Matthew Smith, "Four in Ten Female Millennials Have Been Sent an Unsolicited Penis Photo," *YouGov*, February 15, 2018, https:// yougov.co.uk/topics/politics/articles-reports/2018/02/16/four-ten -female-millennials-been-sent-dick-pic.

24. Flora Oswald, Alex Lopes, Kaylee Skoda, Cassandra L. Hesse and Cory L. Pedersen, "I'll Show You Mine So You'll Show Me Yours: Motivations and Personality Variables in Photographic Exhibitionism," *The Journal of Sex Research* 57, no. 5 (2020): 597–609, doi: 10.1080 /00224499.2019.1639036.

25. National Institute of Justice, "Overview of Stalking," October 24, 2007, https://nij.ojp.gov/topics/articles/overview-stalking.

26. Robin Young and Kalyani Saxena, "Domestic Abusers Are Weaponizing Apps and In-Home Devices to Monitor, Intimidate Victims," WBUR (Boston NPR station), November 27, 2019, https://www.wbur .org/hereandnow/2019/11/27/domestic-abuse-apps-home-devices.

27. For example, see Carrie Goldberg with Jeannine Amber, *Nobody's Victim: Fighting Psychos, Stalkers, Pervs and Trolls* (New York: Plume Press, 2019), and organizations like the SBT Project Foundation, at https://sbtproject.com, and FightCyberstalking.org, at https://www.fightcyberstalking.org/about-fcs/.

28. Gilad Edelman, "Sacred Commandment False Idol," *Wired*, June 2021, 32–47.

29. For more on how deepfakes are created, listen to Kai Rysdall, "Deep thoughts about deepfakes," on *Make Me Smart with Kai and Molly*. Podcast audio, September 3, 2019. https://www.marketplace .org/shows/make-me-smart-with-kai-and-molly/deep-thoughts -about-deepfakes.

30. Clifford Anderson, "Empathy in an Age of Deepfakes," *Cursor_ Zeitschrift für Explorative Theologie*, 2021. https://cursor.pubpub.org /pub/anderson-empathy-deepfakes.

31. Kates Nussman Ellis Farhi & Earle, LLP, "Deepfake Pornography—Risks and Legal Implications," *Firm News*, May 11, 2021. https:// www.katesnussman.com/blog/2021/05/deepfake-pornography-risk s-and-legal-implications/.

32. Anderson, "Empathy in an Age of Deepfakes."

33. For an argument related to using network technology principles to combat use of similar affordances that enable online sex trafficking, see Mitali M. Thakor and Danah Boyd, "Networked Trafficking: Reflections on Technology and the Anti-Trafficking Movement,"

Dialectical Anthropology 37 (2013): 277–90, https://doi.org/10.1007
/s10624-012-9286-6.

34. See chapter 14 in Theodore W. Jennings Jr., *An Ethic of Queer Sex: Principles and Improvisations* (Chicago: Exploration Press, 2013).

35. Ellison, *Making Love Just*, 77.

36. Andrew J. Bauman, *The Sexually Healthy Man: Essays on Spirituality, Sexuality & Restoration* (Las Vegas: No Press, 2021), xiii.

37. Ellison, *Making Love Just*, 81.

38. Elisabeth Vasko, *Beyond Apathy: A Theology for Bystanders* (Minneapolis: Fortress Press, 2015), 8.

39. See FaithTrust Institute, http://www.faithtrustinstitute.org. I am a consultant with FTI, developing curriculum and leading training sessions.

40. West, "Ending Gender Violence," 199.

Chapter 4

1. Second Life is an online virtual world. Users (or residents) interact with one another using avatars. The graphics include everyday setups, like neighborhoods, cities, homes, and so on. There is a complete economy within the Second Life community. You can even get pregnant and give birth to a baby avatar. Nudity and sexual behaviors are allowed only in private rooms or designated clubs. See https://secondlife.com/.

2. Oculus gaming requires a VR headset and hand controls that provide multisensory immersion for first-person experiences. See https://www.oculus.com/.

3. Neil McArthur and Markie Twist, "For the Love of Technology! Sex Robots and Virtual Reality," *The Conversation*, February 6, 2019, https://theconversation.com/for-the-love-of-technology-sex-robot s-and-virtual-reality-110690.

4. Mark Hill, "The Rise of Unapologetically Erotic LGBTQ+ Games," *Wired*, March 4, 2021, https://www.wired.com/story/indie -lgbtq-games-sex/.

5. Ernest Cline, *Ready Player One* (New York: Ballantine, 2011), and Ernest Cline, *Ready Player Two* (New York: Ballantine, 2020).

6. Edward Evans Bailey, "Virtual Reality Sex Is Coming Soon to a Headset Near You," *The Conversation*, May 17, 2016, https://thecon versation.com/virtual-reality-sex-is-coming-soon-to-a-headset -near-you-57563.

7. Justin Ariel Bailey, "The Body in Cyberspace: Lanier, Merleau-Ponty, and the Norms of Embodiment," *Christian Scholar's Review* 45, no. 3 (2016): 222.

8. Bonnie J. Miller-McLemore, "Embodied Knowing, Embodied Theology: What Happened to the Body," *Pastoral Psychology* 62 (2013): 748.

9. On the other hand, embodiment is more than these social categories. Yet, following the wisdom of theologian Bonnie Miller-McLemore when speaking about embodiment, I resist a direct correlation to the physical body because it "can lead to reductionist claims about biological determinism" or a "return to some kind of arrested determinism that ignores culture or presumes or implies that our bodies are our destinies." Miller-McLemore, "Embodied Knowing, Embodied Theology," 744.

10. James Nelson, *Between Two Gardens: Reflections on Sexuality and Religious Experience* (New York: Pilgrim Press, 1983), 21.

11. Melanie L. Harris, *Ecowomanism: African American Women and Earth-Honoring Faiths* (Maryknoll, NY: Orbis Books, 2017). Harris demonstrates the interrelatedness of all of creation by showing how systemic forms of domination falsely separate parts of creation, causing ecological degradation and harm, especially to the most vulnerable.

12. Kate Ott, "Digital Spiritual Embodiment: Power, Difference, and Interdependence," *Cursor_ Zeitschrift für Explorative Theologie* 3 (2019): https://cursor.pubpub.org/pub/3lsdep9t.

13. Donna Haraway, "A Cyborg Manifesto: Science, Technology, and Socialist-Feminism in the Late Twentieth Century," in *Philosophy of Technology: The Technological Condition; An Anthology*, ed. Robert C. Scharff and Val Dusek (Malden, MA: Blackwell Publishing, 2011), 429–66.

14. Two examples of this argument are Philip Butler, *Black Transhuman Liberation Theology: Technology and Spirituality* (New York: Bloomsbury, 2019), and John Dyer, *From the Garden to the City: The Redeeming and Corrupting Power of Technology* (Grand Rapids: Kregel, 2011).

15. See Donna Haraway, *When Species Meet* (Minneapolis: University of Minnesota Press, 2008); Donna Haraway, *Staying with the Trouble: Making Kin in the Chthulucene* (Durham: Duke University Press, 2016); Rosi Braidotti, *The Posthuman* (Malden, MA: Polity Press, 2013); N. Katherine Hayles, *How We Became Posthuman: Virtual Bodies in Cybernetics, Literature, and Informatics* (Chicago: University of Chicago Press, 1999); and Jasbir Puar, *The Right to Maim: Debility, Capacity, Disability* (Durham: Duke University Press, 2017). Those in theological disability studies engage this conversation with increasing creative and inclusive theological possibilities, such as Michella Voss Roberts, *Body Parts: A Theological Anthropology* (Minneapolis: Fortress Press, 2017); Deborah Beth Creamer, *Disability and Christian Theology: Embodied Limits and Constructive Possibilities* (Oxford: Oxford University Press, 2009); Sharon Betcher, *Spirit and the Obligation of Social Flesh* (New York: Fordham University Press, 2014).

16. Kutter Callaway, "Interace Is Reality," in *The HTML of Cruciform Love: Toward a Theology of the Internet*, ed. John Frederick and Eric Lewellen (Eugene, OR: Pickwick, 2019), 24.

17. Callaway, "Interace Is Reality," 24.

18. See Brent D. Laytham, *iPod, YouTube, Wii Play: Theological Engagements with Entertainment* (Eugene, OR: Cascade, 2012) and Bruce Lunceford, "The Body and the Sacred in the Digital Age: Thoughts on Posthuman Sexuality," *Theology and Sexuality* 15, no. 1 (2009): 77–96. Lunceford says that sacredness comes only from humanity's recognition or interaction with the self and other humans, which reduces it to a fleshly exchange. He argues that to transcend or leave the body behind is antisocial, saying, "If sacred experience lies in the removal of mediation, an increase in mediation can only lead one further from the sacred" (94). He believes all humans want to be one flesh or known in an in-person sexual encounter even though he admits that

even those are mediated. This line of argument negates embodied affective and cognitive dimensions of sexuality, while giving preference to the physical acts of sexual behaviors, which has been a long-held Christian bias.

19. Callaway, "Interace Is Reality," 36.

20. *Black Mirror*, season 3, episode 4, "San Junipero," directed by Owen Harris, written by Charlie Brooker, aired October 21, 2016, Netflix.

21. *Black Mirror*, season 5, episode 1, "Striking Vipers," directed by Owen Harris, written by Charlie Brooker, aired June 5, 2019, Netflix.

22. See Marc Loths, "Future Love—VR and the Future of Human Relationships and Sexuality," in *Digital Love: Romance and Sexuality in Games*, ed. Heidi McDonald (Boca Raton: A K Peters/CRC Press, 2017), https://doi.org/10.1201/9781315118444; and McArthur and Twist, "For the Love of Technology!"

23. Rebecca Heilweil, "Elon Musk Is One Step Closer to Connecting a Computer to Your Brain: Neuralink Has Demonstrated a Prototype of Its Brain-Machine Interface That Currently Works in Pigs," *Vox*, August 28, 2020, https://www.vox.com/recode/2020/8/28/21404802/elon-musk-neuralink-brain-machine-interface-research.

24. I have explored this *Black Mirror* episode in depth, and some of this description originates in another published article from the point of view of afrofuturist queer theologies. See Kate Ott, "Purifying Dirty Computers: Cyborgs, Sex, Christ, and Otherness," *Cursor_ Zeitschrift für Explorative Theologie*, April 13, 2021, https://cursor.pub pub.org/pub/ott-purifying-dirty-computers.

25. For analysis of the intersections of US Christianity with racial, class, gender, and economic constructions of the family, see Rosemary Radford Ruether, *Christianity and the Making of the Modern Family* (Boston: Beacon, 2001).

26. Elizabeth Landau, "The Found Community, and Then Love, in Online Games," *Wired*, December 1, 2012, https://www.wired.com/story/love-community-mmorpg-online-gaming/.

27. Pawel Tacikowski, Jens Fust, and H. Henrik Ehrsson, "Fluidity

of Gender Identity Induced by Illusory Body-Sex Change," *Scientific Reports* 10 (2020): https://doi.org/10.1038/s41598-020-71467-z.

28. Tacikowski, Fust, and Ehrsson, "Fluidity of Gender Identity Induced by Illusory Body-Sex Change."

29. Angelina Aleksandrovich and Leonardo Mariano Gomes, "Shared Multisensory Sexual Arousal in Virtual Reality (VR) Environments," *Paladyn, Journal of Behavioral Robotics* 11, no. 1 (2020): 379–89, https://doi.org/10.1515/pjbr-2020-0018.

30. Jason Parham, "The Strange, Subtle Matter of ASMR Erotica," *Wired*, January 24, 2020, https://www.wired.com/story/asmr-erotica/.

31. Bailey, "Virtual Reality Sex Is Coming Soon to a Headset Near You."

32. Emily Witt, *Future Sex—a New Kind of Free Love* (New York: Farrar, Straus & Giroux, 2016).

33. "Tinder, Teledildonics & Sex in 2020: Salon Series with Tina Horn, Dr. Tammy Nelson, Mal Harrison, and Sue-Yae Johnson," March 2018 *Future of Sex* podcast, and "O6: Can Teledildonics + VR = Human Intimacy? (Featuring Dame Products, Kiiroo and Bado-inkVR)," January 2017 *Future of Sex* podcast. For further information about the podcast, visit https://www.futureofsex.org/podcast.

34. Theodore W. Jennings Jr., *An Ethic of Queer Sex: Principles and Improvisations* (Chicago: Exploration Press, 2013), 200.

35. Scott A. Midson, "From *imago dei* to Social Media: Computers, Companions and Communities," in *Love, Technology, and Theology*, ed. Scott A. Midson (London: T&T Clark, 2020), 150.

36. Midson, "From *imago dei* to Social Media," 155.

37. Midson, "From *imago dei* to Social Media," 156.

38. David Gunkel and Debra Hawhee, "Virtual Alterity and the Reformatting of Ethics," *Journal of Mass Media Ethics* 18 (2003): 173–93.

Chapter 5

1. For the purposes of this chapter, a robot is a mechanical device that performs tasks based on a command or programmed instructions. That definition might include lots of items around the house

like a digital coffee machine or the washing machine. However, those are more commonly considered automation rather than robots. A device is automated when it can follow through on a programmed task often in a limited environment. The coffee machine won't move and it certainly won't learn anything as it brews your coffee. There is some debate over whether a machine's ability to sense even without movement makes it a robot. So maybe the high-end, advanced coffee machine that self-measures water, senses the weight of coffee grounds, and calculates a grounds-to-water ratio is a robot. Virtual assistants are also not robots since they generally do not assess their surroundings to complete tasks or have movement. See Ben Dickson, "Why AI Assistants Can't Be Robots (For Now)," Tech Talks, April 8, 2019, https://bdtechtalks.com/2019/04/08/ai-digital-assistants-alexa -robot/.

2. "Tinder, Teledildonics & Sex in 2020: Salon Series with Tina Horn, Dr. Tammy Nelson, Mal Harrison, and Sue-Yae Johnson," March 2018 *Future of Sex* podcast, and "O6: Can Teledildonics + VR = Human Intimacy? (Featuring Dame Products, Kiiroo and Bado-inkVR)," January 2017 *Future of Sex* podcast. For further information about the podcast, visit https://www.futureofsex.org/podcast.

3. Jack Derwin, "Sex Tech Is Now a $30 Billion Industry. An Expert Talked Us through 10 Gadgets with Very Different Uses," *Business Insider Australia*, February 12, 2020, https://www.businessinsider .com.au/sex-technology-gadgets-wrap-ai-robots-apps-2020-2.

4. James McBride, "Robotic Bodies and the Kairos of Humanoid Theologies," *Sophia* 58 (2019): 663–76, https://doi.org/10.1007/s11841 -017-0628-3.

5. John Danaher, "Should We Be Thinking about Robot Sex?," in *Robot Sex: Social and Ethical Implications*, ed. John Danaher and Neil McArthur (Cambridge: MIT Press, 2017), 4–5.

6. For a full history and current review of available sex robots, see Kate Devlin, *Turned On: Science, Sex and Robots* (New York: Bloomsbury, 2020).

7. Julie Carpenter studies how the medium of technology affects humans' interactions with it. Her initial research was on relationships

with robots in military settings. In her research, which she extends now to sex robots, she notes that human sociocultural acceptance of robots requires a process of cultural familiarity that takes place over a span of time. Media and myth development is part of the shift toward acceptance, thus the importance of considering religious myths related to creation and significant media portrayals of sex robots. Overall, she brings together a variety of models of human adaptation to and acceptance of robots in support of her thesis. See Julia Carpenter, "Deus Sex Machina: Loving Robot Sex Workers and the Allure of an Insincere Kiss," in Danaher and McArthur, *Robot Sex*, 261–88.

8. Danaher, "Should We Be Thinking about Robot Sex?," 6.

9. "Sex Dolls: How AI-Enabled Real Dolls Are Coming to Life (Feat. Matt McMullen of RealDoll)," March 2017 *Future of Sex* podcast.

10. See Kate Ott, *Christian Ethics in a Digital Society* (New York: Routledge, 2019), chapter 1.

11. Francis X. Shen, "Sex Robots Are Raising Hard Questions," *Fast Company*, February 18, 2019, https://www.fastcompany.com /90308471/sex-robots-are-raising-hard-questions.

12. See Campaign Against Sex Robots, https://campaignagainst sexrobots.org/.

13. John Danaher, Brian Earp, and Anders Sandberg, "Should We Campaign against Sex Robots?," in Danaher and McArthur, *Robot Sex*, 47–72.

14. Litska Strikwerda, "Legal and Moral Implications of Child Sex Robots," in Danaher and McArthur, *Robot Sex*, 133–52.

15. "Curbing Realistic Exploitative Electronic Pedophilic Robots Act, H.R.4655-CREEPER Act of 2017, 115th Congress (2017–2018).

16. For a fuller treatment of this topic and additional research, see chapter 9 in Kate Devlin, *Turned On: Science, Sex and Robots* (New York: Bloomsbury, 2020), and Yuefang Zhou, "Preventive Strategies for Pedophilia and the Potential Role of Robots: Open Workshop Discussion," in *AI Love You: Developments in Human-Robot Intimate Relationships*, ed. Yuefang Zhou and Martin H. Fischer (Cham: Springer Nature Switzerland AG, 2019).

17. _Westworld: The Maze_ (season one), creators Jonathan Nolan and Lisa Joy, HBO, 2016. The series is based on the 1973 film written and directed by Michael Crichton.

18. Carpenter, "Deus Sex Machina."

19. Lela Moore, "'My Best Source of Comfort': Adults with Stuffed Animals Describe All the Feels," _New York Times_, December 20, 2018, https://www.nytimes.com/2018/12/20/reader-center/adults-with -stuffed-animals.html.

20. See ParoRobots, http://www.parorobots.com/.

21. See Softbank Robotics, https://www.softbankrobotics.com /emea/en/pepper.

22. See McBride, "Robotic Bodies and the Kairos of Humanoid Theologies." Additionally, the movie _Lars and the Real Girl_ captures very genuine discussions about and community acceptance of Lars's sex doll (not a robot) companion as a church member.

23. Devlin, _Turned On_, 55.

24. For an in-depth philosophical account of the claim of reciprocity related to intention without consciousness, see Viktor Kewenig, "Intentionality but Not Consciousness: Reconsidering Robot Love," in Zhou and Fischer, _AI Love You_. For more on the ethical design features, see Devlin, _Turned On_, 147–49; Allie Gemmill, "One of the World's Most Famous Sex Robots Can Now Revoke Her Consent," _Dazed_, June 24, 2018, https://www.dazeddigital.com/science-tech/ar ticle/40478/1/samantha-the-sex-robot-can-refuse-sex-if-not-in-the -mood; and, Alex Williams, "Do You Take This Robot . . . ," _New York Times_, January 19, 2019, https://www.nytimes.com/2019/01/19/style /sex-robots.html.

25. See Devlin, _Turned On_, 164–81.

26. Giuseppe Lugano, Martin Hudák, Matúš Ivančo, and Tomáš Loveček, "From the Mind to the Cloud: Personal Data in the Age of the Internet of Things," in Zhou and Fischer, _AI Love You_.

27. David Levy, _Love and Sex with Robots_ (New York: HarperCollins, 2007).

28. "Woman Builds Inmoov a 3D Printed Robot Proceeds to Fall in

Love and Now Wants to Marry It," *TECHEBLOG*, December 27, 2016, https://www.techeblog.com/woman-builds-inmoov-a-3d-printed -robot-proceeds-to-fall-in-love-and-now-wants-to-marry-it/.

29. Sherry Turkle, *Alone Together: Why We Expect More from Technology and Less from Each Other* (New York: Basic Books, 2011).

30. Devlin, *Turned On*, 159–62.

31. This characterization of treatment of sex dolls by their owners is not universal but predominant. A number of factors might contribute to this, including the high cost of the doll, customization to the owner's preferences, and the connection most owners create with their doll. As a counter to this argument, media and critics of sex dolls like to point to an incident at a 2017 technology conference where Samantha, Sergi Santos's robot doll, was on display. It was repeatedly touched and prodded to make its sensors respond, just as any display model in a store might be used. Media exploited the story, saying the sex doll was raped and violated. Santos tried to clarify in the media that this was not the case, but the clarification never received the same press coverage until Kate Devlin clarified the record in her book. She writes, "This was a robot on display as a trade show prototype and people were being told they could touch it. Tens of thousands of visitors, having expressly been given permission to press the flesh of a delicate doll, had acted on the invitation. . . . Putting a spin of intentional sexual violence on this is downright misleading." Devlin, *Turned On*, 229.

32. "Living with Sex Dolls (Feat. Davecat)," November 2020 *Future of Sex* podcast, https://www.futureofsex.org/podcast.

33. Noreen Herzfeld, "Religious Perspectives on Sex with Robots," in Danaher and McArthur, *Robot Sex*, 99.

34. Steve Petersen, "Is It Good for Them Too? Ethical Concern for the Sexbots," in Danaher and McArthur, *Robot Sex*, 160.

35. William F. Schulz and Sushma Raman, *The Coming Good Society: Why New Realities Demand New Rights* (Boston: Harvard University Press, 2020), section III.

36. One reason for denying rights to nonhuman animals or en-

tities is often a "category mistake," or assuming a right has to apply to all people or things. For example, a cisgender sexed male doesn't need a right to abortion and a dog doesn't need the right to vote. Situational or circumstantial application of rights is needed with animals and certain AI, like autonomous weapons, but not a robovac (Schulz and Raman, *The Coming Good Society*, 169, 189).

37. Schulz and Raman, *The Coming Good Society*, 57.

38. I do not subscribe to the idea that materials like pornography or sex robot use damage whole societies or directly cause violence against women and girls, for example. Banning pornography and sex robots will not end rape and harassment. Yet regulation of certain material depictions and experiences of sexual violence should be curbed to promote a social ethos against violence.

39. Herzfeld, "Religious Perspectives on Sex with Robots," 95.

40. Kino Coursey, Susan Pirzchalski, Matt McMullen, Guile Lindroth, and Yuri Furuushi, "Living with Harmony: A Personal Companion System by RealbotixTM," in Zhou and Fischer, *AI Love You*, 86–88.

41. Jana M. Bennett, "We Do Not Know How to Love: Observations on Theology, Technology, and Disability," *Journal of Moral Theology* 4, no. 1 (2015): 90–110.

42. "Loving Persons. Activity and Passivity in Romantic Love," in *Love and Its Objects: What Can We Care For?*, ed. Christian Maurer, Tony Milligan, and Kamila Pacovská (Houndmills: Palgrave Macmillan, 2014), 41–55.

43. Thomas Jay Oord, "Can Technologies Promote Overall Well-Being? Questions about Love for Machine-Oriented Societies," in Midson, *Love, Technology, and Theology*, 127–42.

44. Devlin, *Turned On*, 205.

45. Scott A. Midson, "From *imago dei* to Social Media: Computers, Companions and Communities," in Midson, *Love, Technology, and Theology*, 155.

46. Midson, "From *imago dei* to Social Media," 156.

47. Anne Foerst, "Loving Robots? Let Yet Another Stranger In," in Midson, *Love, Technology, and Theology*, 70.

48. Foerst, "Loving Robots," 72.

49. Foerst, "Loving Robots," 73.

50. Devlin, *Turned On*, chapter 10.

51. John Danaher, "Building Better Sex Robots: Lessons from Feminist Pornography," in Zhou and Fischer, *AI Love You*, 143.

52. Danaher, "Building Better Sex Robots," 143.

Selected Bibliography

Sexual Ethics and Christian Theology

Bauman, Andrew J. *The Sexually Healthy Man: Essays on Spirituality, Sexuality & Restoration.* Las Vegas: No Press, 2021.

Brownson, James V. *Bible, Gender, Sexuality: Reframing the Church's Debate on Same-Sex Relationships.* Grand Rapids: Eerdmans, 2013.

Carr, David M. *The Erotic Word: Sexuality, Spirituality, and the Bible.* New York: Oxford University Press, 2003.

De La Torre, Miguel. *Liberating Sexuality: Justice between the Sheets.* St. Louis: Chalice Press, 2016.

———, ed. *Out of the Shadows into the Light: Christianity and Homosexuality.* St. Louis: Chalice Press, 2009.

Ellison, Marvin M. *Making Love Just: Sexual Ethics for Perplexing Times.* Minneapolis: Fortress, 2012.

Farley, Margaret. *Just Love: A Framework for Christian Sexual Ethics.* New York: Continuum, 2006.

Fortune, Marie M. *Sexual Violence: The Sin Revisited.* Cleveland: Pilgrim Press, 2005.

Gary, Sally. *Affirming: A Memoir of Faith, Sexuality, and Staying in the Church.* Grand Rapids: Eerdmans, 2021.

Jennings, Theodore W. *An Ethic of Queer Sex: Principles and Improvisations.* Chicago: Exploration Press, 2013.

Keen, Karen R. *Scripture, Ethics, and the Possibility of Same-Sex Relationships*. Grand Rapids: Eerdmans, 2018.

Klein, Linda Kay. *Pure: Inside the Evangelical Movement That Shamed a Generation of Young Women and How I Broke Free*. New York: Atria Paperback, 2018.

Moultrie, Monique. *Passionate and Pious: Religious Media and Black Women's Sexuality*. Durham: Duke University Press, 2017.

Traina, Cristina L. H. *Erotic Attunement: Parenthood and the Ethics of Sensuality between Unequals*. Chicago: University of Chicago Press, 2011.

Vasko, Elisabeth. *Beyond Apathy: A Bystander Theology*. Minneapolis: Fortress, 2015.

Weems, Renita. *What Matters Most: Ten Lessons in Living Passionately from the Song of Solomon*. New York: Walk Worthy Press, 2004.

West, Traci C. *Disruptive Christian Ethics: When Racism and Women's Lives Matter*. Louisville: Westminster John Knox, 2006.

———. *Solidarity and Defiant Spirituality: Africana Lessons on Religion, Racism, and Ending Gender Violence*. New York: New York University Press, 2019.

Digital Technology, Theology, and Ethics

Butler, Philip. *Black Transhuman Liberation Theology: Technology and Spirituality*. New York: Bloomsbury, 2019.

Dyer, John. *From the Garden to the City: The Redeeming and Corrupting Power of Technology*. 2nd ed. Grand Rapids: Kregel, 2022.

Frederick, John, and Eric Lewellen, eds. *The HTML of Cruciform Love: Toward a Theology of the Internet*. Eugene, OR: Pickwick, 2019.

Midson, Scott A., ed. *Love, Technology, and Theology*. London: T&T Clark, 2020.

Ott, Kate. *Christian Ethics for a Digital Society*. New York: Rowman & Littlefield, 2019.

Sexuality, Media, and Technology

Berg, Heather. *Porn Work: Sex, Labor, and Late Capitalism.* Chapel Hill: University of North Carolina Press, 2021.

Cottle, Michelle. "The Adultery Arms Race." *The Atlantic.* November 2014. https://www.theatlantic.com/magazine/archive/2014/11/the-adultery-arms-race/380794/.

Danaher, John, and Neil McArthur, eds. *Robot Sex: Social and Ethical Implications.* Cambridge: MIT Press, 2017.

Devlin, Kate. *Turned On: Science, Sex and Robots.* New York: Bloomsbury, 2020.

Farris, Nicole, D'Lane R. Compton, and Andrea P. Herrera, eds. *Gender, Sexuality and Race in the Digital Age.* Cham: Springer, 2020.

Goldberg, Carrie, with Jeannine Amber. *Nobody's Victim: Fighting Psychos, Stalkers, Pervs and Trolls.* New York: Plume Press, 2019.

Hill, Mark. "The Rise of Unapologetically Erotic LGBTQ+ Games." *Wired.* March 4, 2021. https://www.wired.com/story/indie-lgbtq-games-sex/.

Julian, Kate. "The Sex Recession." *The Atlantic*, December 2018, 78–94.

Rosin, Hanna. "Why Kids Sext." *The Atlantic*, November 2014. https://www.theatlantic.com/magazine/archive/2014/11/why-kids-sext/380798/.

Ross, Karen. *The Handbook of Gender, Sex, and Media.* Oxford: Wiley & Sons Ltd, 2012.

Slater, Dan. *A Million First Dates: Solving the Puzzle of Online Dating.* New York: Penguin, 2014.

Tolentino, Jia. "The Rage of the Incels: Incels Aren't Really Looking for Sex. They're Looking for Absolute Male Supremacy." *The New Yorker.* May 15, 2018. https://www.newyorker.com/culture/cultural-comment/the-rage-of-the-incels.

Witt, Emily. *Future Sex—a New Kind of Free Love.* New York: Farrar, Straus & Giroux, 2016.

Zhou, Yuefang, and Martin H. Fischer, eds. *AI Love You: Developments in Human-Robot Intimate Relationships*. Cham: Springer Nature Switzerland AG, 2019.

Public Research Studies

Merritt, Jonathan. "Pornography: A Christian Crisis or Overblown Issue?" Religion News Service, January 20, 2016. https://reli gionnews.com/2016/01/20/christians-pornography-problem/.

"The Virtues and Downsides of Online Dating." Pew Research Center, February 6, 2020. https://www.pewresearch.org/internet/2020 /02/06/the-virtues-and-downsides-of-online-dating/.

Vogels, Emily A. "The State of Online Harassment." Pew Research Center, January 13, 2021. https://www.pewresearch.org/in ternet/2021/01/13/the-state-of-online-harassment/.

Pop Culture

Fiction

Cline, Ernest. *Ready Player One*. New York: Ballantine, 2011.
———. *Ready Player Two*. New York: Ballantine, 2020.
Littman, Sarah Darer. *Deepfake*. New York: Scholastic, 2020.
Weinstein, Alexander. *Children of the New World*. New York: Picador, 2016.
———. *Universal Love: Stories*. New York: Holt, 2020.

Podcasts

Future of Sex. https://www.futureofsex.org/podcast.
Make Me Smart: Tech, the Economy, and Culture. https://www.mar ketplace.org/shows/make-me-smart-with-kai-and-molly/.
Restorative Faith, Season 2: Sexuality. https://www.restorativefaith .org/season-2.
Sex Out Loud with Tristan Taormino. https://sexoutloud.libsyn.com.

Television and Movies

A.I. Artificial Intelligence, 2001
Black Mirror, Netflix Series, 2001–2019
The Circle, Netflix Series, 2020–2021
Ex Machina, 2014
Her, 2013
Lars and the Real Girl, 2007
Romantic Chorus—an Animated Documentary, 2001, producer distrib-
    uted at https://vimeo.com/ondemand/romanticchorus
Westworld, HBO Series, 2016–2020

Youth and Young Adult Sexuality Resources

Corinna, Heather. *S.E.X., the Second Edition: The All-You-Need-to-Know*
    Progressive Sexuality Guide to Get You through High School and
    College. Boston: DeCapo Press, 2016.
———. "Sexuality: WTF Is It Anyway." http://www.scarleteen.com
    /article/bodies/sexuality_wtf_is_it_anyway.
Ott, Kate. "The Dating Game: How to Survive the Roller Coaster."
    TheThoughtfulChristian.com, https://www.thethoughtful
    christian.com/Products/TC5058/the-dating-game.aspx.
———. "Defining Pornography." Youth Study. TheThoughtfulChris
    tian.com, https://www.thethoughtfulchristian.com/Products
    /TC5208/defining-pornography.aspx.
———. "Forming a Healthy Relationship." TheThoughtfulChristian
    .com. https://www.thethoughtfulchristian.com/Products
    /TC5060/forming-a-healthy-relationship.aspx.
———. "Responding to Violence in Jesus' Time and Now." TheThought
    fulChristian.com, https://www.thethoughtfulchristian.com
    /Products/TC5107/responding-to-violence-in-jesus-time-and
    -now.aspx.
———. *Sex + Faith: Talking to Your Child from Birth to Adolescence.* Lou-
    isville: Westminster John Knox, 2013.

———. "What Digital Story Are You Writing?" TheThoughtfulChris
tian.com, https://www.thethoughtfulchristian.com/Prod
ucts/TC5098/what-digital-story-are-you-writing.aspx.

Index